D0848491

THE PLASTICS AGE

THE PLASTICS AGE

FROM BAKELITE TO BEANBAGS AND BEYOND

Edited by Penny Sparke

The Overlook Press
Woodstock, New York

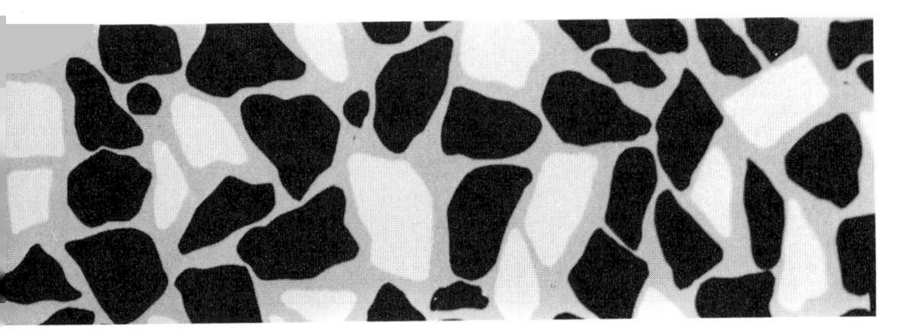

First published in 1993 by
The Overlook press
Lewis Hollow Road
Woodstock, New York 12498

Library of Congress Cataloging-in-Publication Data

The Plastics age : from bakelite to beanbags and
beyond /Penny Sparke

p. cm.
Includes bibliographical references and index
1. Plastics.
2. Design, Industrial, I. Sparke,Penny.

TP1122.P554 1993
668.4–dc20

92--35039
CIP

ISBN : 0-87951-471-X (HC)
ISBN : 0-87951-488-4 (PB)

Designed by Bernard Higton
Printed in Singapore by Craft Print
Picture research by Julia Engelhardt
We wish to acknowledge permission
to reprint material from existing
publications:
Robert Friedel
M.I.T. Press
Penguin Books
Unwin Hyman
C.I.P.I.A.
New Society
The Estate of Roland Barthes and
the translator, Annette Lavers
Editions Callimard
Ezio Manzini
Rizzoli International Publications Inc.

CONTENTS

PART ONE: PLASTICS PRE-HISTORY 1860-1914

PART TWO: PLASTICS AND MODERNITY 1915-1960

PART THREE: PLASTICS AND POST-MODERNITY 1961-1990

A range of 'tortoiseshell' celluloid goods, including a powder bowl, a cigarette case, a glove stretcher and boxes, 1890s-1920s.

Introduction

ON THE MEANINGS OF PLASTICS IN THE TWENTIETH-CENTURY

'Plastics have exploited their formable qualities not so much in the direction of fulfilling technical and constructional needs as in the expression of different images.'
Ezio Manzini[1]

The Italian historian of technology, Ezio Manzini, has written that the material culture that constitutes our everyday environment is a result of a compromise between what is 'possible' and what is 'thinkable' – in other words, between technology in its most advanced state and the 'ideas' that are generally acceptable by the mass of the populace, ie; culture. This formula can be used to explain the nature of many of the man-made 'things' which surround us and facilitate our daily lives.

Those plastic products which are consumed and used by the mass market are explicable in these terms, particularly when (as is the case in this book) it is their design, rather than their technological make-up, that is under primary consideration. Whatever the 'state of the art' of the research into the chemical components of that ever increasingly complex family of materials, these products still have to filter through a cultural net which requires that they meet the symbolic demands of consumers and users. The decisions relating to these requirements are, for the most part, made not by technologists but by designers and they depend upon the nature of the cultural status quo as well as upon current design ideals and preferences.

Where utility dominates other factors – in the areas of, say, space and medical research – the technical input is everything, whereas the plastics used for making a washing-up bowl, a piece of furniture or a pair of shoes are chosen as much for their aesthetic appeal and symbolic significance as for their economic and technological advantages.

The cultural meanings that plastics have acquired over the course of this century are, however, varied and complex. Created, in the first place, as substitutes for luxury materials which were in increasing demand and diminishing supply in the second half of the nineteenth century, first celluloid and later the synthetic plastics were developed as cheaper, more available alternatives. Simultaneously however, they also served as an essential adjunct to a number of the new technological innovations of the era – photography, the early film industry and the electrical industry in particular. Thus plastics' reputation was formed, in the early days, within the twin realms of the expansion of the middle-class market and the technological advances of the 'modern world': As such they earned respect and admiration.

Only gradually, with the socio-cultural changes of the first half of this century and the emergence of so-called 'mass culture' at the end of this era, did plastics fall from grace and become transformed, in the eyes of society at large, into 'cheap and nasty' substitutes. The demise in their reputation was influenced by a number of factors, among them the large-scale manufacture of plastics and plastic products – dominated by American companies such as DuPont – which rid them of any lingering associations with luxury and quality that they might have retained from an earlier period, and the growing availability of such products to a mass market.

After the Second World War plastic products became associated with the concepts of 'inauthenticity', cheapness', 'low quality' and 'bad taste'. In contrast to the traditional 'craft' materials – clay, stone etc – they were downgraded and seen as essentially 'inferior' – a reputation which they have been trying, with the help of designers, to throw off ever since. The most lasting association was with the idea of 'falseness', or 'artificiality': This carried with it overt moral overtones inherited from the Modern Movement in architecture and design which was formed in the last years of the last century and the early years of the twentieth-century, and required that man-made objects respect the dictum of 'truth to materials'. The problem with considering plastics in these terms is that there is no one truth but rather a plurality of 'possibilities', in Manzini's sense of the term. Judged by this yardstick plastics inevitably fall short of the necessary requirements and present us, as a result, with a cultural dilemma.

An advertisement for Bakelite Ltd. *Modern Plastics* USA March 1934.

In the late 1950s and 1960s a group of designers sought a way out of this impasse. Dismayed by the sinking reputation of these exciting and, above all, modern materials, a number of furniture designers in Italy and product designers in a number of European countries, set out to find an 'authentic form' for plastic products which conformed to the Modern Movement criteria. Their researches led them into a renewed discussion of its principles, considered now in the light of the technological necessities of plastics manufacturing. A number of what have since become 'classic' plastic products emerged in these years – among them Vico Magistretti's 'Selene' dining chair; Dieter Rams' 'cool' designs for the housings of many of the Braun company's electrical and electronic goods; and the sophisticated, small domestic objects created by designers such as Gaby Schreiber and David Harman-Powell for a number of British manufacturers. They proved (or so it seemed at the time) that plastics had found a 'natural' minimal

Plastic flowers made by Arthur Miller, 1951.

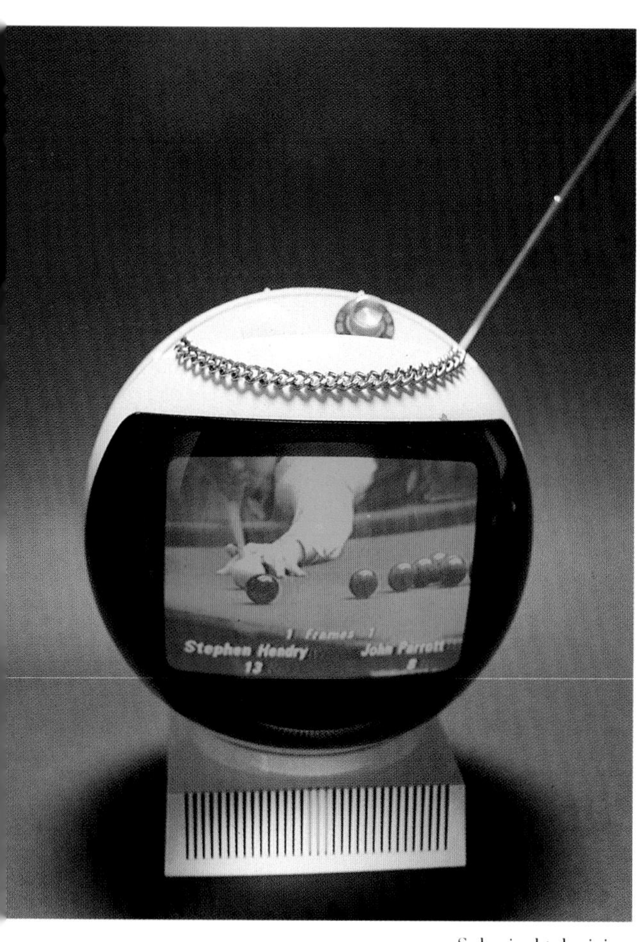

Spherical television,
1960s.

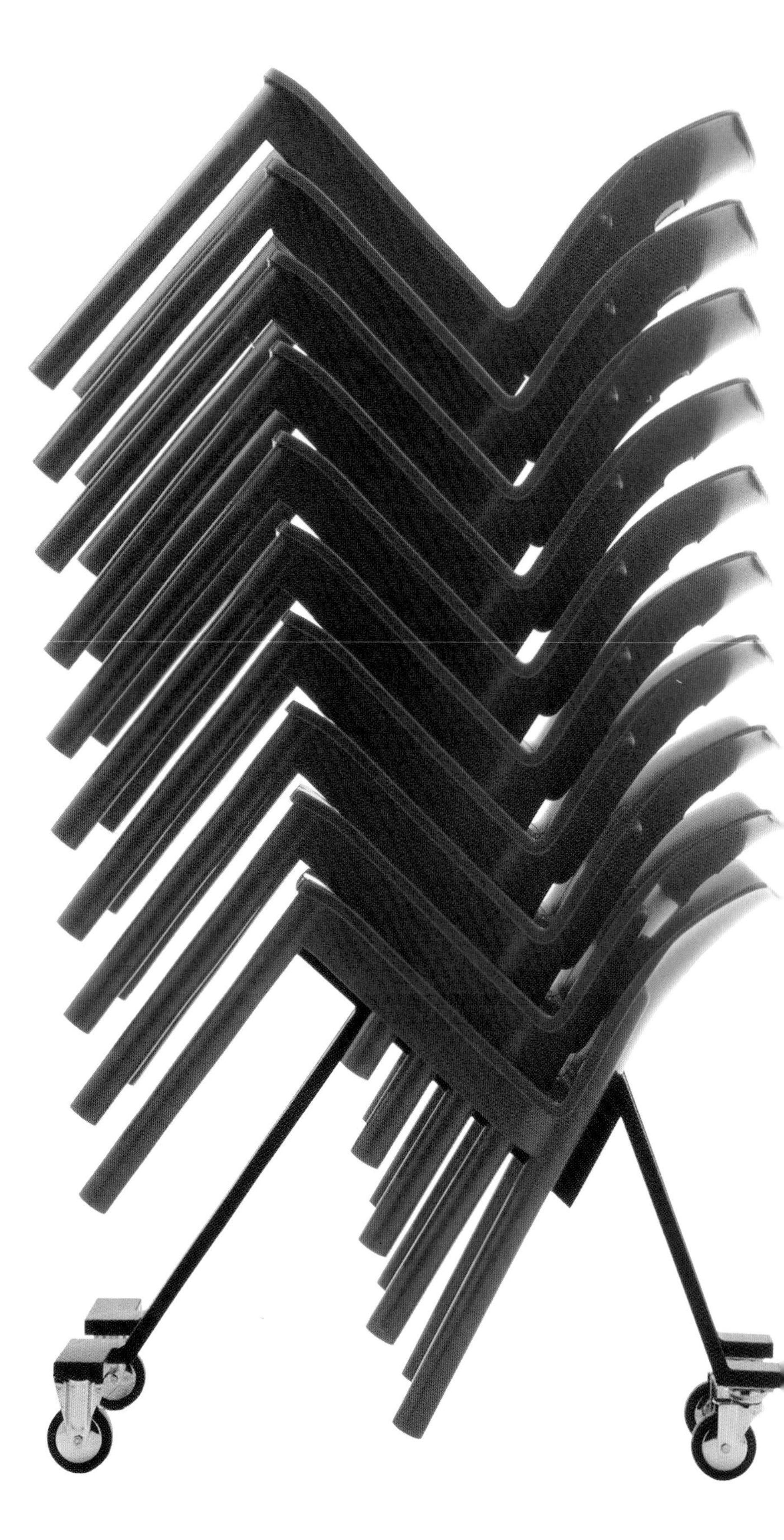

Anna Castelli Ferrieri.
Stacking chairs, model
4870, manufactured
by Kartell SpA. Italy
1986.

aesthetic for themselves, characterised by sleek surfaces, organic curves and bright, clear colours.

While there are direct relationships between the moulding process and the flowing forms of these products the idea that there was one appropriate aesthetic for all plastic products was illusory and, as the subsequent Pop and Post-Modern movements went on show, with the ever-expanding technological possibilities made available to plastics manufacturers and designers an ever-widening spectrum of aesthetic alternatives also became viable. The full realisation of this fact required, however, a shift in the cultural quo before it could become a reality. The general move away from the ideals of the Modern Movement which took place from the late 1960s onwards proved to be the necessary catalyst, allowing plastics, once again, to become respected on their own terms – ie; as materials which could perform a number of different cultural roles simultaneously.

With the emphasis moving away from the moral inadequacies of the aesthetic and symbolic significance of plastic products, their critics began, instead, to focus on their ecological implications, in particular on the fundamental problem of their disposability. Once again plastics became, for a time, the helpless victims of hostile criticism, considered inferior to the 'natural' materials which could return into the earth from whence they came, without harming the environment.

While the ecological outcry of the 1970s has, by now, abated somewhat, it has not entirely disappeared and many 'defences' of plastics stress the fact that research is being carried out in this area, and that recycling plastics is now a reality. At the same time, however, the reputation of plastics as a cultural force has been much restored by their expanding role within areas such as medical and space research. As symbols of advanced technology which can ameliorate man's condition they are widely seen now as vital ingredients of 'modernity'.

We remain, ultimately, however, ambivalent about plastics and their complex cultural meanings. They are capable of denoting both decadence and progress simultaneously. Their Jeckyll and Hyde character permits them to reflect various aspects of contemporary culture and their unique position at the intersection between high culture and mass culture renders them highly potent materials capable of mirroring many of the values within contemporary society.

In the end we return to Manzini's formula about the 'thinkable' and the 'possible'. As both become increasingly complex so there is a growing need for conscious control over the way in which they relate to each other. That is the job of the designer: to manipulate the meanings of materials as they enter social and cultural life. This anthology charts some of the ways in which plastics have acquired their cultural meanings through this century. Through the words of critics and historians who have focused their attention on this fascinating subject we learn how plastics have come to mean what they do today and why they still occupy such an uneasy place within contemporary cultural life.

Penny Sparke

Notes

1. E. Manzini. 'And of Plastics?' in *Domus*, 666 (Nov. 1985), p.54.

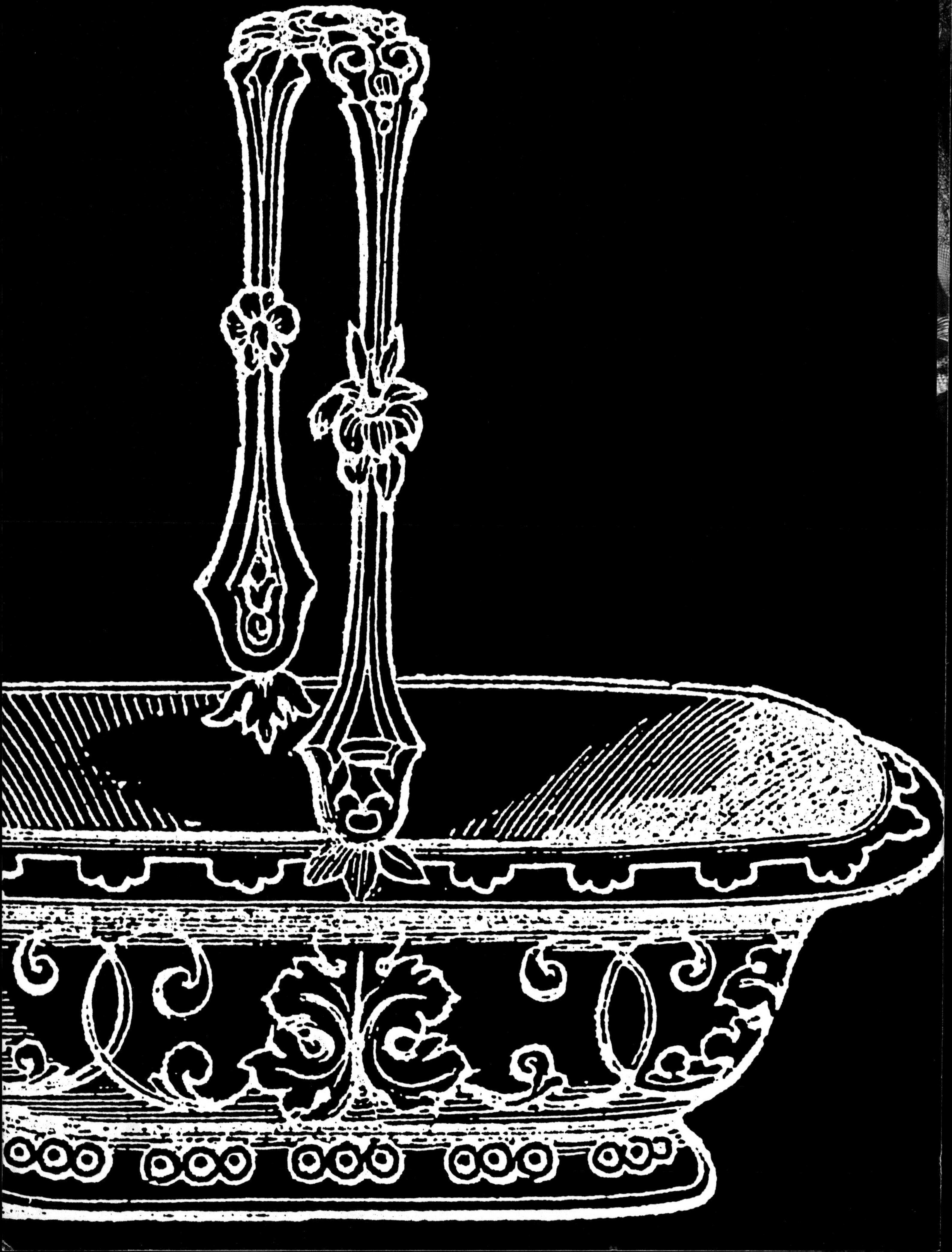

PART ONE

PLASTICS PRE-HISTORY 1860-1914

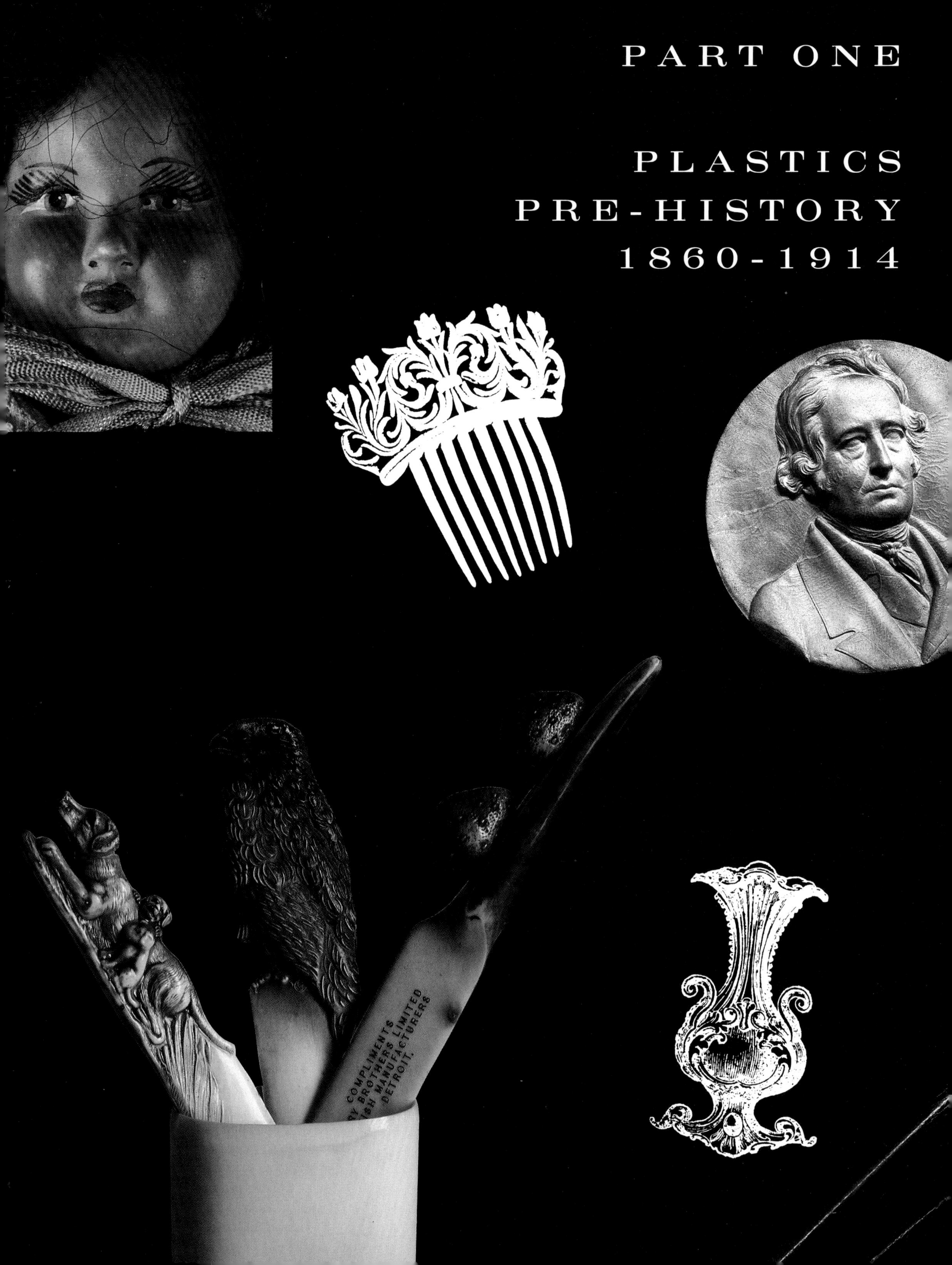

INTRODUCTION

A comb polisher. Oyonnax, France. Late nineteenth-century.

The emergence of the 'natural' and the 'semi-synthetic' plastics in the second half of the nineteenth century was the result of both the technological explosions and the dramatic cultural changes that occurred during this period: Both supply and demand were enormously powerful forces in influencing the dramatic changes in the material environment of the late nineteenth-century industrialised world and any account of early plastics which does not consider both perspectives is incomplete.

In understanding both why and how the early plastics emerged it is vital, therefore, to consider both their technological make-up and their general role within society and culture. To this end the two essays in this short first section of the book focus on these areas. Together they provide a useful introduction to the story of plastics in the twentieth century.

Susan Mossman from the Science Museum in London concentrates on 'The Technology of Early Plastics'. She lists the most significant materials in turn and describes their means of manufacture and their applications thereby demonstrating the richness of invention in this area during this period.

The American social historian, Robert Friedel, on the other hand, focuses in his essay, 'The First Plastic' (reprinted from *American Heritage of Invention and Technology* Summer 1987, Vol. 3 No. 1), on the socio-cultural context within which celluloid was made and sold in the USA in the second half of the nineteenth century. Less specifically concerned with the actual composition of the newly invented material he concentrates instead on the broader significance of its penetration into the market-place. He stresses celluloid's role as a cheaper, susbstitute material for older

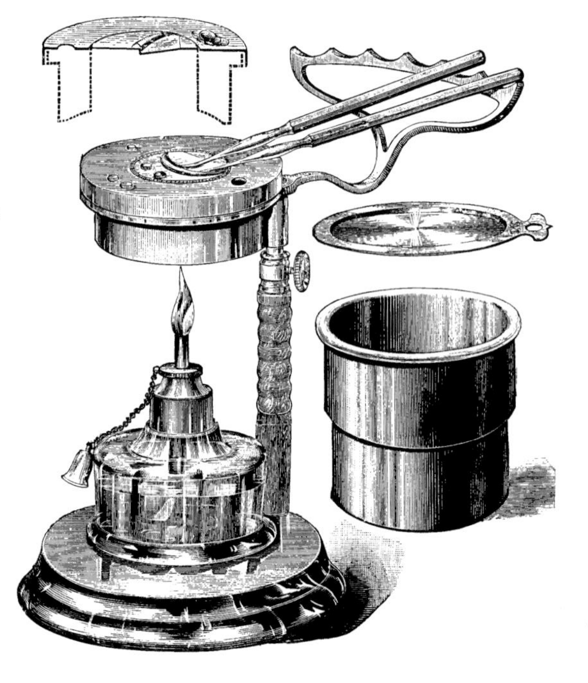

Illustration from a dental catalogue. 1886.

A range of women's articles made of celluloid simulating horn and tortoiseshell. USA Late nineteenth-century.

ones which were becoming increasingly expensive and points out that it owed its success to the marketing exercise which emphasised this function.

Whether or not we accept Friedel's thesis that Hyatt was the 'inventor' of celluloid, rather than the British argument which places Alexander Parkes in the important role, is not really at issue here. What is vital, however, is the realisation that without a ready market no invention can be successful. In Friedel's words, celluloid was treated as 'a raw material for artistry and ornament' and it was this essential adaptability to the prevailing taste of the day that allowed it to penetrate the environment in a significant manner. As 'an instrument of social mobility' it played an important role in representing the values of American middle-class society at that time. In Manzini's words, therefore, the 'possible' coincided perfectly with the 'thinkable'.

Penny Sparke

GUTTA PERCHA

THE TECHNOLOGY OF EARLY PLASTICS

Susan Mossman

Many people think of Bakelite when they think of early plastics. There is also a tendency to believe that plastics are a twentieth century phenomenon. Certainly, this is true of the phenolic polymers which are totally man-made and were invented and first patented in 1907 by Leo Baekeland. They were later developed as a moulding material from which the first totally synthetic plastics were manufactured under the name Bakelite. This essay will concentrate on their predecessors: the natural and semi-synthetic plastics. The semi-synthetics came into existence around the middle of the nineteenth century whereas the natural plastics have been known for many centuries, some going back into the mists of antiquity.

The evidence for the early technology of plastics comes in many forms: both written and artefactual. The patents submitted by the inventors themselves provide an invaluable insight into the production techniques used to manufacture early plastics. In certain cases, these men have also recorded their discoveries in the form of monographs or published lectures which also help to give us an insight into the conditions (often adverse) under which they worked.

Natural Plastics

Natural plastics are organic substances which when heated can be softened and then formed in moulds. They have a very long history, some dating back to the ancient Egyptians; they are also numerous and include amber, horn, wax, bitumen, shellac, and gutta percha.

Applications of gutta percha. 1850.

The techniques used to make items from horn, especially pressing, had a strong influence on the methods subsequently developed to manufacture plastics. The methods utilised to mould natural plastics, for example heating and pressure moulding, were very successfully adapted to fabricate the semi-synthetic and later the entirely synthetic plastics. Naturally there were some innovations in the production equipment, notably in the case of the machinery used to produce vulcanite and celluloid.

Gutta percha is a dark-brown substance obtained from the Palaquium tree (which is indigenous to Malaysia). During the latter part of the nineteenth century gutta percha had many applications, ranging from elaborately moulded picture frames to ornate inkstands, and even including ear trumpets; these decorative mouldings were produced by placing hot gutta percha in cold moulds. In 1845, Bewley adapted existing extrusion equipment for plastics. This extrusion process was applied to the manufacture of gutta percha tubes, and made possible, in 1850, the production of gutta percha sheathing for the first submarine telegraph cable. Gutta percha was selected for this purpose since it possesses excellent

Rubber ball from a Peruvian child's grave. Circa 1650.

insulating properties. Objects made of gutta percha are fragile, and become brittle and cracked with time, or may not survive at all; however, we know that it was made into an astonishing variety of different shapes due to the number of pattern books and trade catalogues which still remain.

Shellac is the name given to the secretion of the 'lac' beetle and was particularly popular in the latter half of the nineteenth century. This material could be mixed with a number of different fillers, such as wood flour, and placed in a heated mould which, on cooling, released the final object. Shellac had the particular quality of being able to reproduce very fine moulded detail, and it was often used to make objects ornamented with elaborately moulded scenes, especially hand mirrors and other cosmetic items. In addition it was a very suitable medium for the earliest gramophone records in the 1880s. An American, Samuel Peck, patented a method in 1854 of mixing:

'gum shellac and woody fibers or other suitable fibrous material dyed to the color that may be required and ground with the shellac and betweeen hot rollers so as to be converted into a mass which when heated becomes plastic so that it can be pressed into a mold or between dies . . .'[1]

Peck used this 'natural plastic' to make the so-called 'Union Cases' which had ornate decorated surfaces and were designed to hold daguerreotypes or ambrotypes (the predecessors of today's framed family photographs).

Semi-synthetic Plastics

The earliest semi-synthetic plastics were invented in the mid-nineteenth century.

The substances known as vulacanite or ebonite in Britain and as 'hard rubber' in the USA deserve discussion since they are materials made from a natural material,

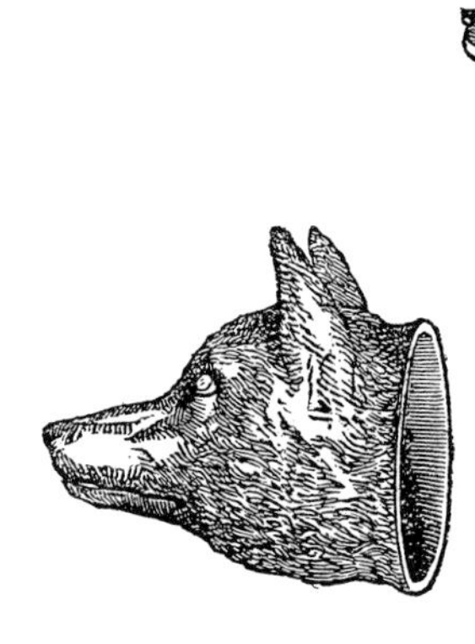

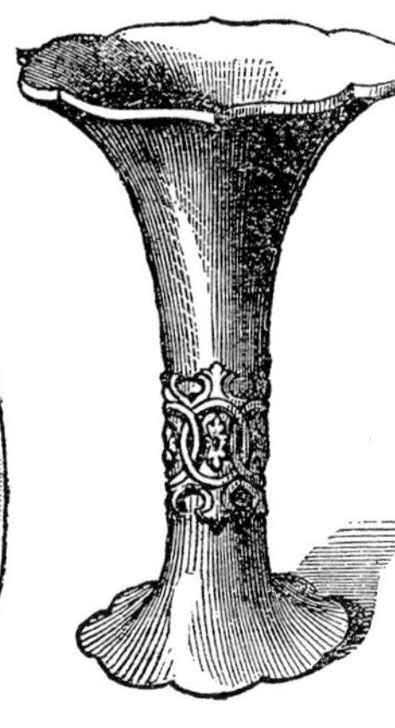

Page from gutta percha company's prospectus. 1851.

Plaque of Thomas Hancock (1786-1865) made of white vulcanised rubber.

rubber, which has been processed, being altered by the addition of sulphur under controlled conditions; they are among the earliest so-called semi-synthetic plastics.

The method called vulcanisation which was used to produce vulcanite in the nineteenth century is well known. Fortunately, Thomas Hancock, the British inventor of the vulcanisation process in 1844, published the story of his invention. He wrote:

'I now saw my course straight before me; my experiments had shown that the rubber and sulphur must first be blended . . . Sulphur could be blended by rollers and masticators; the blended material could be reduced to a state of solution by any of the usual solvents, or to a state of dough, and spread on cloth by the machinery of my previous patents for waterproofing, or on sized cambric for sheets. Blocks of the masticated compound could be pressed in moulds and cut into sheets, and into every variety of form and size that the new uses of the 'changed' rubber might require; and last, though not least, ordinary cut sheets of pure rubber in the form they came from the knife, or made up into other forms, could be immersed in a sulphur bath, and changed to any required degree, from the softest and most elastic up to a state of hardness similar to horn, and capable of being wrought with carpenters' tools, or turned like ivory and ebony in a lathe.' [2]

Hancock also informs us that the name of the process is due to his friend Mr Brockedon who suggested the term 'vulcanization' after the god, Vulcan, who employed both sulphur and heat.[3]

Hancock in Britain[4] and Goodyear in the USA[5] perfected the vulcanisation process. This led to the production of a wide range of objects made of 'hard' or 'vulcanised' rubber which will be very familiar to many. Tyre manufacture was a very practical

Rubber cartoon circa 1830. Charles Macintosh was the inventor of the original Macintosh, a waterproof coat.

exploitation of this material's particular qualities, but it was also employed for imitation jet jewellery, Vesta match boxes and other decorative items. In contrast, during the nineteenth and early twentieth centuries, vulcanite played an important role in the manufacture of dentures; it was mixed with red and white pigments to produce a pink substance which imitated gums. The disadvantage of vulcanite was that it was usually black in colour, which rather limited its use for ornamental items. The periods of mourning for Prince Albert and then Queen Victoria must have benefited the producers of vulcanite jewellery, since this substance made a very satisfactory substitute for the more expensive jet.

Casein, another semi-synthetic plastic, was also very popular in the late nineteenth century; the first patents being issued in Germany and America in 1885 and 1886 respectively. Casein was patented as Galalith in Germany, and as 'Erinoid' in Britain. Casein is made from milk curds which are dried, mixed with water to produce a dough, which is extruded and then treated with formaldehyde producing a material which resembles horn or bone. Casein can be moulded into a variety of shapes, moreover a dazzling array of colours can be created by surface dyeing or the addition of pigments, as is revealed by the dye recipe books for casein which have survived, notably in the Science Museum, London. Casein was used to make a great number of ornamental articles including jewellery and decorative boxes, and will be well known to many in the form of buttons. The major disadvantage of casein is that it absorbs water very easily and hence distorts. Casein lost favour with the advent of many new plastics after the Second World War.

What of Celluloid? The story of the invention of its predecessor, Parkesine, by Alexander Parkes is a familiar one; notable for the failure of Parkes to capitalise on his invention. Parkes produced cellulose nitrate from nitric acid, sulphuric acid and cellulose (which was obtained from sources such as cotton) and mixed it with vegetable oils and small proportions of organic solvents; the result was a dough which could be moulded or pressed into sheets and which Parkes named Parkesine after himself. Parkesine could be used to make a wide variety of objects from dough which was either first heated to soften it and then pressed into moulds, or, alternatively, hand-

Hancock's experimental masticator.

Plaque moulded from Parkesine. Circa 1855.

carved and then inlaid with mother-of-pearl or metal wire. Parkes successfully exhibited his invention at the Great International Exhibition of 1862, but failed to make a commercial success of the Parkesine Company.

Daniel Spill then attempted to make money out of this product (renamed Xylonite) which had great limitations due to its flammability, although its production process was relatively simple. Spill set up the Xylonite Company in 1869, and produced a range of decorative objects made of Xylonite and Ivoride (a white form of Xylonite).[6] These included hand mirrors, fancy combs, and knife handles. Spill failed in his attempts to make a success of the Xylonite Company, hampered by lengthy and ultimately unsuccessful legal wrangles with his American competitor, Hyatt, and he died an embittered man.

It was left to the American, John Wesley Hyatt, with the help of his brother Isaiah, to carry Parkes' invention forward in the attempt to make a more stable plastic. Hyatt succeeded when he concentrated on camphor and discovered that it made an excellent solvent and plasticiser for cellulose nitrate when used in precise quantities. In his own words:

'*First*: The idea of combining with the nitrocellulose only the exact or approximate amount of solvent required for a solid solution. This required a nearly perfect mechanical mixture before very much solvent action could take place.'

Hyatt christened his invention Celluloid and set up the highly successful Celluloid Company.

Hyatt played an important role in the development of specific plastics manufacturing machinery. The Hyatts' engineer, Charles Burroughs, designed specialised tools and machinery for celluloid production. Hyatt mentions 'The stuffing machine process'.[7] The 'stuffing machine', patented in 1872, produced celluloid 'in the form of a bar, sheet or stick'[8] which was then machine finished.[9] This machine is considered to be the predecessor of the modern injection moulding machine.

Celluloid's flammability remained a serious drawback and Hyatt even received letters of complaint about exploding celluloid billiard balls. The search was on for a less flammable plastic. One solution was cellulose acetate, but it was not produced in a mouldable powder form until 1929.

The production of celluloid and similar products proliferated worldwide. It is fortuitous that a celluloid factory still survives in Britain, at Wardle Storeys in

Brantham, Essex, the home of the old Xylonite Company, and it is still possible to visit and see how celluloid is produced, by methods which have remained almost unchanged since the factory was built in 1887. In addition, some of the original recipe books for the production of Xylonite can be consulted at Ipswich Public Records Office. One is reminded of cooking on a large scale. Unlike the finely regulated plastics industry of today, at the turn of the century the production of celluloid very much depended on the operator's experience. Sometimes this was not sufficient, and the machinery would seize up with a mixture which was too thick; precious time would have to be spent on unblocking the machinery. Many of the recipes, for example, to produce many of the tortoiseshell and pearlised effects, were never even written down but were jealously guarded secrets handed on by word of mouth. In the heyday of the Xylonite production at this factory, in the late nineteenth and early part of the twentieth century, many different tortoiseshell patterns were produced; now the range is limited to three.

In conclusion, plastics before the First World War consisted of semi-synthetic materials, namely celluloid, casein and vulcanite. The processes used to manufacture them were frequently adapted from those which were already in use for shaping natural plastics such as horn. Specialised machinery was developed for the production of celluloid in particular, and this machinery exerted great influence on the development of many modern plastics production methods.

Susan Mossman
The Science Museum, London

Above: Objects made of Xylonite and Ivoride, manufactured by D. Spill and Company. Circa 1870.

Right: Block press used to make cellulose nitrate sheets. Tangyes Limited Birmingham. 1948.

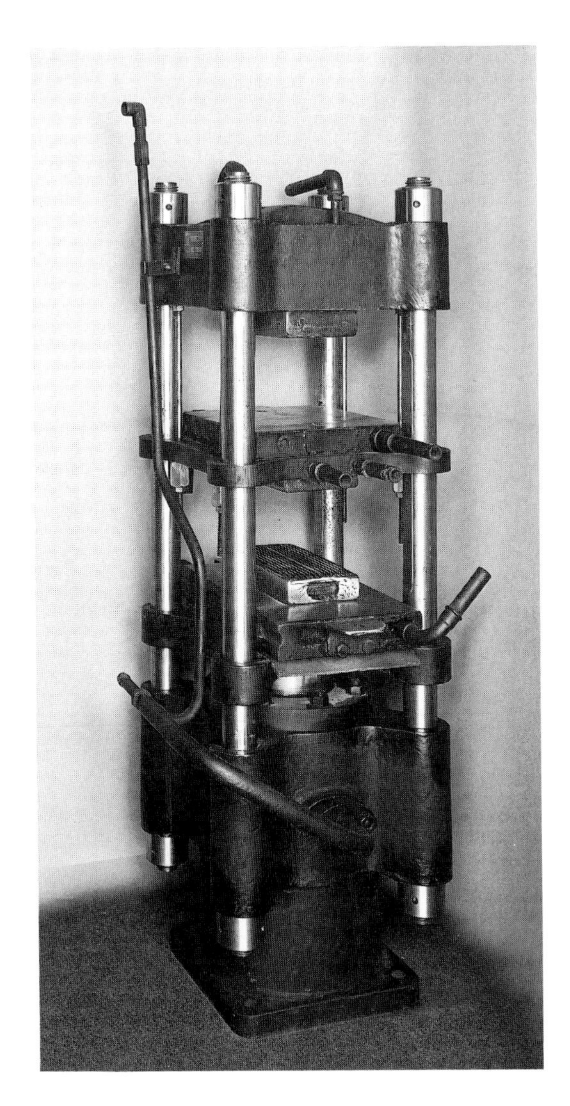

Acknowledgements
I wish to thank Dr Robert Bud for encouraging me to write this essay and Heather Mayfield, Jan Metcalfe, John Ratcliffe, Dr Derek Robinson and Colin Williamson for their advice and comments.

Notes
1. US Patent 11758, 1854. lines 20-7
2. T. Hancock. *Personal Narrative of the Origin and Progress of the Caoutchouc or India-Rubber Manufacture in England.* p.104-5
3. Ibid, p. 107
4. British Patent 9952 1843
5. US Patent 3633, 1844
6. M. Kaufman *The First Century of Plastics.* p.30
7. R. Friedel *Men, Materials, and Ideas: A History of celluloid*, PhD thesis John Hopkins University (submitted 1976). University Microfilms International, Ann Arbor, Michigan 1978, p.44
8. US Patent 133229
9. Kaufman. op. cit., note 6, p.101

NO
TRUMPS
MONTREAL CAN
NO

THE FIRST PLASTIC

Robert Friedel

Plastic: In an everyday context the word connotes the cheap, the disposable, even the flimsy. However much we may appreciate the obvious usefulness of plastics in a whole host of applications and even their technical superiority at sometimes crucial points (from airplane windows to football helmets), most of us are never far from a sense of skepticism or disdain as we encounter this enormous and varied class of materials in our daily lives. This attitude accounts for our surprise, then, when we see century-old examples of the first plastic and are faced not with the mundanely utilitarian but with artistic efforts previously reserved for precious and semiprecious substances.

This first plastic, celluloid, was invented in 1869 by an Albany, New York printer, John Wesley Hyatt. Over the next several decades celluloid and its imitators came to be used and sold in a variety of forms. While often the material was presented as an inexpensive replacement for such substances as ivory, tortoiseshell, and amber, it generally was treated not as a cheap, utilitarian stuff but rather as a raw material for artistry and ornament. As such it became a vehicle for spreading a taste for luxury and the 'finer things' among the rapidly growing middle class of late-nineteenth-century America. In other words, this first plastic was an instrument of social mobility – a social mobility of consumption that would come to be one of the distinguishing (and, to the rest of the world, the most enviable) features of American society during the last hundred years.

This truth is brought home not so much by what was said about celluloid as by the objects themselves. Brought together in collections, the combs, brushes, letter openers, dolls, tableware, and dozens of other articles formed from celluloid speak eloquently of its meaning for its purchasers and owners. In recent years older plastic items have become respectable objects of interest for museums and collectors. A few established institutions, such as the Smithsonian, have notable collections, and there is even movement toward the establishment of a national plastics museum in one of the first centers for the manufacture of the material, Leominster, Massachusetts. But for at least the moment the most impressive collections are probably still in private hands. In the case of celluloid, few, if any, collections match that of Dadie and Norman Perlov in New York City. Their more than seven thousand pieces suggest that the Perlovs have pursued this material with the

'Spider' celluloid comb. Oyonnax, France. Late nineteenth-century.

Above left: Celluloid doll. Late nineteenth-century.

Below left: Celluloid imitating ivory in the form of three trump markers, dice and a box for playing cards. USA Early 1900s.

persistence and dedication associated with only the most serious collectors.

Our first impression upon seeing the Perlov's or any other celluloid collection is likely to be that we have stumbled across a trove of ivory, for not only is the distinctive cream colour to be seen everywhere, but so are, on closer inspection, the minute striations that signify the growth of an elephant's tusk. This ivory form was only one of the many effects that could be convincingly produced in celluloid, but it was overwhelmingly popular. This seems appropriate, for in developing the material, John Wesley Hyatt was seeking a substitute for one of ivory's most distinctive uses, the billiard ball.

According to the oft-told tale, when Hyatt formulated his new material be was after a prize for an ivory replacement offered by a major billiards supply house. Like a number of experimenters before him, he started with a syrupy substance known as collodion. To make collodion, which was used as a kind of liquid bandage or as a carrier for photosensitive chemicals in some photographic processes, ordinary cotton (or some other vegetable fiber) was first nitrated and then dissolved in alcohol and ether. Cotton can be nitrated simply by being reacted with nitric and sulfuric acids. The resulting nitrocellulose is often called guncotton, and it first received attention as an explosive. Collodion, while highly flammable, is not explosive, but spread thin it dries into a clear, flexible film. Several would-be inventors tried to make collodion yield a thicker and stronger material but without success. It was Hyatt who discovered that nitrocellulose, only slightly moistened with standard solvents, could be mixed under pressure with camphor, an aromatic crystalline gum, to produce a hard, translucent solid.

Some more experimentation on the new material, dubbed celluloid by Hyatt and his brother, Isaiah, showed that it could be easily formed and coloured. Indeed, some spectacular effects could be achieved, yielding not only nearly exact imitations of ivory but also the black-and-tan mottling of tortoiseshell, the shimmer of mother-of-pearl, and the rich golden tone of amber. One thing celluloid did not do, however, was make a good billiard ball. While the plastic resembled ivory in feel and appearance, its density and elasticity were sufficiently different to defeat its original purpose. Nor were the Hyatt brothers especially pleased with the results of their second effort at commercialization – dental plates. The appeal of the smooth material for such a use was obvious: It was made in shades of fleshy pink, and alternatives like hard rubber were notoriously ugly and uncomfortable. But celluloid's tendency to soften in very hot water could be disconcerting to unsuspecting tea drinkers, and its camphor odor (and taste) were hard to eliminate entirely. As is often the case with new materials (or other novel technologies), finding the right markets was a slow and often frustrating process.

During the 1870s and 1880s the celluloid makers slowly introduced their material into a variety of uses, some of them failing and others becoming familiar items of American life. The more technical applications (such as emery wheels or truss pads) tended to be unsuccessful. It rapidly became clear that celluloid's most important virtue was its ability to look and feel like something else –

Celluloid hat-pins, with their stands, adorned with elaborate paintwork. USA Late nineteenth-century.

COMPLIMENTS
BROTHERS LIMITED
MANUFACTURERS
DETROIT.

something more rare and valuable. Added to this ability was the ease with which the material could be cut, moulded, bent, stamped, and otherwise formed into a myriad of shapes. The result was that celluloid came to be made into what merchants called novelties and fancy goods.

These items – buttons, letter openers, hatpins, jewelry boxes, tape measures, thimbles, brushes, mirrors, combs, and the like – were the kinds of things to be found in most ordinary households. They were not particularly important or noteworthy in themselves, but they were the small articles that made up (and still constitute) the paraphernalia of everyday life – things of convenience and pleasure and ornament rather than objects of necessity. Celluloid was rarely the cheapest material for such items, for hard rubber, tin, bone and cardboard, among other substances, were frequently available for the least expensive tastes. The plastic, instead, held a kind of middle ground, neither the cheapest nor the costliest but rather a sensible moderate choice. By choosing celluloid, the consumer could at once announce his or her appreciation of good taste and fine appearance while eschewing ostentation and extravagance. Here, then, was a material for the middle classes.

The manufacturers of celluloid used considerable ingenuity to make their products successfully fill this role. To execute the ivory effect, for example, the producers stacked thin sheets of the plastic, each tinted in slightly varying shades of cream-white, and then used high pressure, heat, and solvents to force the sheets into a solid block. Cutting the block across the grain yielded very convincing striations in the resulting piece. The popular result was shaped into almost every article imaginable, many of them, such as postcards, vases, boxes, and dolls, improbable or impossible in true ivory. The appeal to fashion was heightened by such names as Ivaleur, Ivorine, or French ivory, for in the nineteenth century Parisian styles were regarded as the last word in good taste. Finally, the material lent itself, as ivory never could, to ornamental effects that combined a sense of artistry and design with mass production. The fancy monograms that adorned dresser sets, the complex molded forms that were made into hatpins, letter openers, or toothpicks, the intricate reliefs applied to everything from jewelry boxes to beer foams (paddles that skim the head of a glass of beer) – all testified to the plastic's role as a substance of beauty.

This role came at a price, however. As one can see in looking through the thousands of articles in the Perlovs' collection, celluloid was perceived and treated as a purely imitative material. It was almost always made to look like something else – something more expensive and harder to fabricate. This first plastic, therefore, became known as an imitation, rather than as a wonderful new substance to be exploited for its own good looks. There was no technical reason, after all, for celluloid not to be made into bright, shiny colors, exploiting the kinds of properties we have come to associate with plastics in the twentieth century. It is, in fact, easier and cheaper to make celluloid that way than to imitate ivory or tortoiseshell. Spectacle frames of celluloid, to take just one example, could in the nineteenth century have been made as bold and bright as we make high-

Four decorative celluloid letter-openers. USA Late nineteenth-century.

fashion sunglasses today. They were not. Celluloid was instead used to fill the demand for tortoiseshell frames, a fashion that did not really catch on until after the First World War. If one of our most common images of plastics today is as a cheap imitation, one of the reasons lies in the history of this first plastic.

Celluloid did, of course, come to fulfil some very important non-imitative functions. In the late 1880s a Newark, New Jersey, pastor, Hannibal Goodwin, discovered how to make celluloid into a thin, uniform, transparent film. At about the same time, Henry Reichenback, a chemist working for George Eastman's photographic supply house in Rochester, New York, made a similar discovery and devised a method of producing photographic film from celluloid. This was just what Eastman was looking for to put into his new camera, the Kodak. From the moment of its introduction in the summer of 1889, celluloid film had a tremendous impact on photography, an impact that was to be amplified by its application to motion pictures during the following decade.

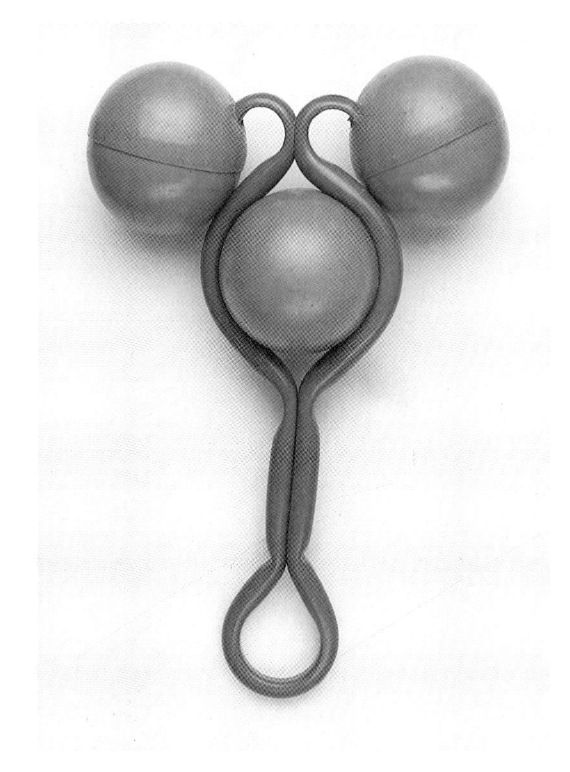

Baby's rattle made from celluloid sheet and tube. Probably France, late nineteenth-century.

Indeed, in the twentieth century the word *celluloid* has been more often used to connote movies than to suggest plastics.

The original makers and fabricators of celluloid did not participate in or gain much from the photographic revolution sparked by their material – the patents and monopoly practices of Eastman saw to that. They instead continued to produce articles appealing to the ever-growing middle class

Celluloid perfume bottles manufactured by Halex. England, late nineteenth-century.

and to an urban working class that was beginning to acquire some of the means and habits of consumption identified with the more affluent. Their products continued to be accessible tokens of good taste and of an appreciation of the finer things, all speaking of the owner's sense of belonging to a world where luxury and beauty and style were within the grasp of everyone.

Celluloid continued to be an important commercial material into the 1930s, but by then most of the old companies had been absorbed by larger, more progressive chemical manufacturers, such as Celanese and Du Pont. More significantly, rival materials had begun to emerge with increasing frequency, particularly as chemists began to build an impressive structure of theory to guide the synthesis of what were now called polymers. The invention of the synthetic resin Bakelite in 1907, by the Belgian-trained chemist Leo Baekeland, may be seen in hindsight as a major first step in this direction, and the rapid adoption by John Hyatt's own company of Bakelite for making billiard balls was an apt symbol of the versatility of the new plastic. The flammability of celluloid was a great incentive to find substitutes for it even in photography and motion pictures, and the somewhat more expensive safety film (cellulose acetate) began displacing celluloid widely as the movie industry grew in size and importance. (The distantly related material cellophane was invented in 1908.)

By the late 1940s celluloid's markets had so shrunk that it was no longer of commercial importance. Today there is very little celluloid produced in the United States, and its applications are largely confined to certain kinds of fuses (utilizing its great flammability), fancy decorations for jukeboxes, and, in a particularly sweet irony, Ping-Pong balls, for which celluloid appears to be as uniquely suited and irreplaceable as ivory billiard balls seemed to be a century ago.

Robert Friedel is the Advanced Research Fellow at the Hagley Museum, in Delaware, Associate Professor in the Department of History at the University of Maryland, and the author of *Pioneer Plastic: The Making & Selling of Celluloid* (University of Wisconsin 1983).

Celluloid wrist-bag. Late nineteenth-century.

PART TWO

PLASTICS AND MODERNITY 1915-1960

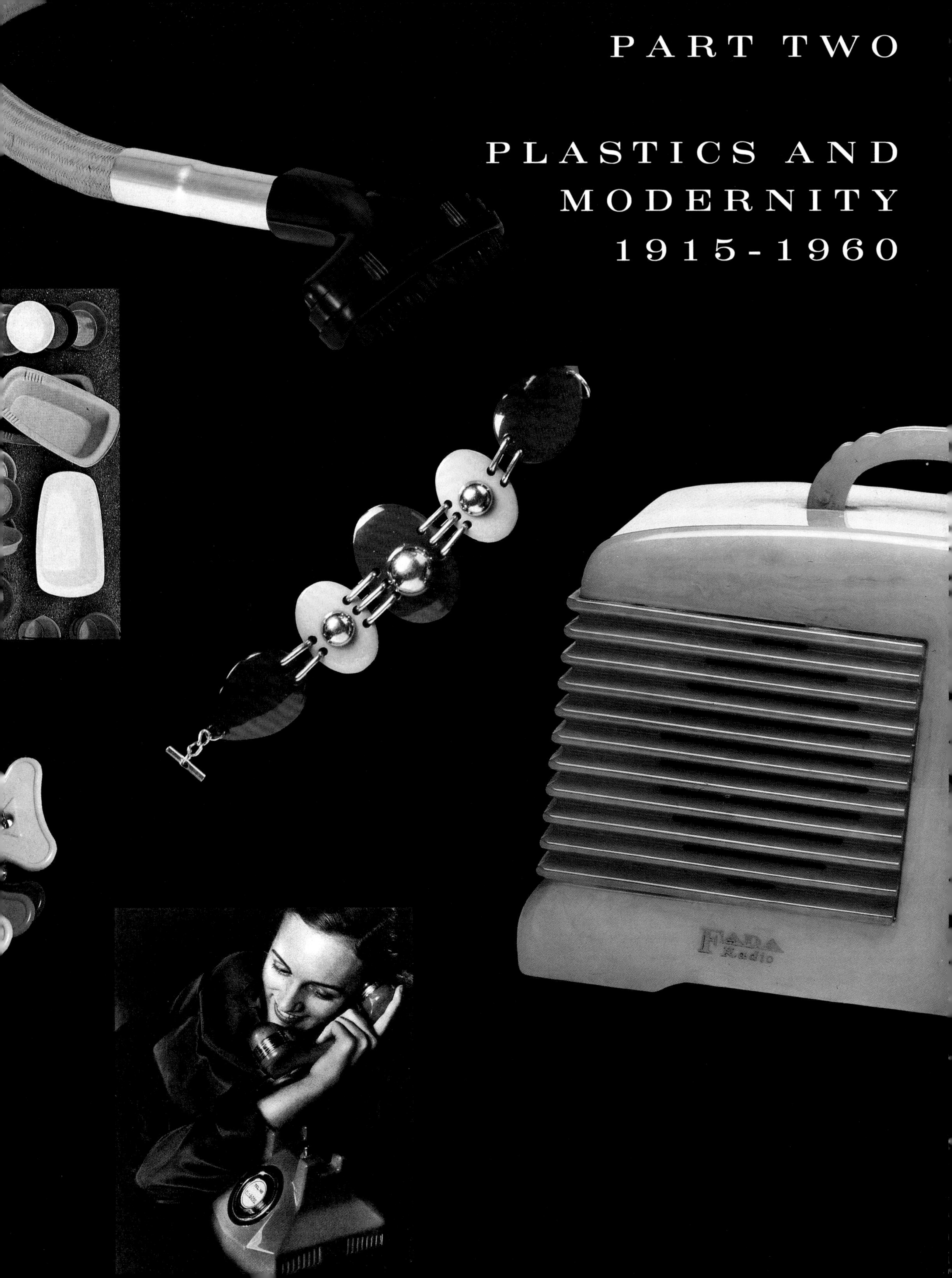

INTRODUCTION

An analysis of the 'ideology of technology', continues to motivate much of the writing in Section Two of this anthology which focuses on the period associated with the first and second generations of Modernism in design. Once again an ambivalent attitute towards plastics emerges in this period: They are seen, particularly in the USA, as the new liberating materials of the 'modern age' while, a little later, especially in Britain, they become symbols of vulgarity, pastiche and inauthenticity. Subsequently, however, through the hands of a number of manufacturers and designers in Europe they are reconstituted, in this period, as the materials of modernity.

The essays in this section chart a path which moves through optimism to pessimism and back again to optimism in relation to the role of plastics within modern life, and it covers a geographical route which moves from the USA, to Britain and on to Europe, particularly Italy. This reflects the chronological pattern of the technological evolution in plastics manufacture which was strongly influenced by developments which took place on American soil. After the Second World War Europe was quick to emulate the example set by the USA.

Jeffrey Meikle's essay, 'Plastics in the American Machine Age 1920-1950' (condensed by the author from an earlier piece entitled 'Plastic, Material of a Thousand Uses', published in *Imagining Tomorrow: History, Technology and the American Future*, edited by Joseph J. Corn and published by the MIT Press in 1988), focuses on the changing cultural identity of

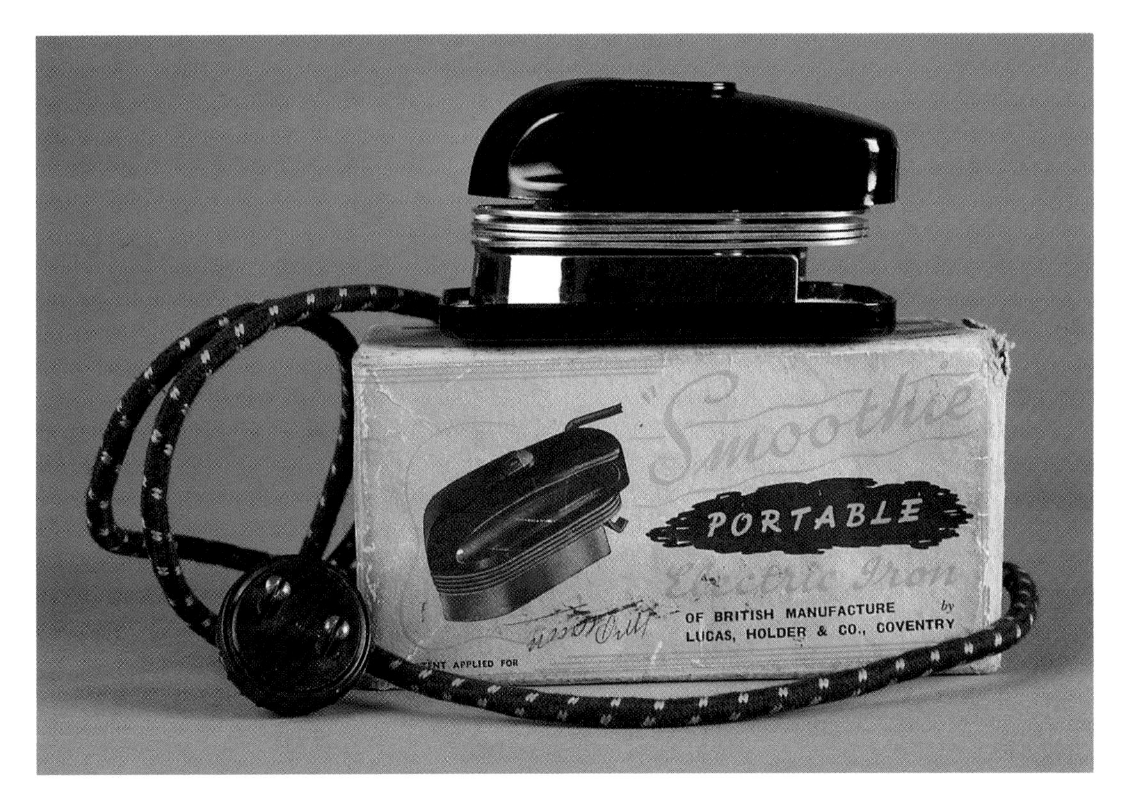

Portable iron. Black phenolic and chromed steel. USA 1930s.

Serge Chermayeff. Radio AC-86. Table model in bakelite. 1935. England.

plastic products in the USA in the 'heroic' period of their evolution, between the two World Wars. It is, he writes, 'the story of how plastics first gained and lost their utopian significance'. He begins by providing an account of the way in which plastics acquired their image of 'miracle materials of the futuristic machine age' with the aid of the new professional of those years – the industrial designer. It was he, claims Meikle, who acted as a catalyst in the shift from plastics being seen as substitute materials to their becoming symbols of modernity. The designer created an image of the future, epitomised by the style dubbed 'streamlining'.

While Meikle's thesis ends with a brief account of the demise of plastics utopianism in the USA, brought about by their war-time role as 'ersatz' materials and the post-war rekindling of the desire for 'genuine' materials, such as wood and leather, V.E. Yarsley and E.G. Couzens, writing in Britain in their little book entitled, simply, *Plastics*, published by Pelican in 1941, (this excerpt is taken from the final chapter, 'Plastics and the Future') sustains the mood that Meikle had identified in pre-Second World War America, by enthusing about the role that they saw for plastic products across the Atlantic. This short piece from their evocative text captures the spirit of optimism that underpins their account and shows the extent to which plastics utopianism penetrated British thinking about the subject in the years during and just after the war. The exaggerated nature of their futuristic vision underlines their commitment to a world in which plastics dominated practically every aspect of the everyday life of their imaginary 'Plastic Man'. Most of their predictions, which seemed mere science-fiction projections at the time, have, of course, subsequently become the stuff of reality.

Writing in his book *Plastics and Industrial Design*, published by Allen and Unwin four years later, the British design critic and historian, John Gloag, reiterates Yarsley and Couzens' enthusiasm for the new materials

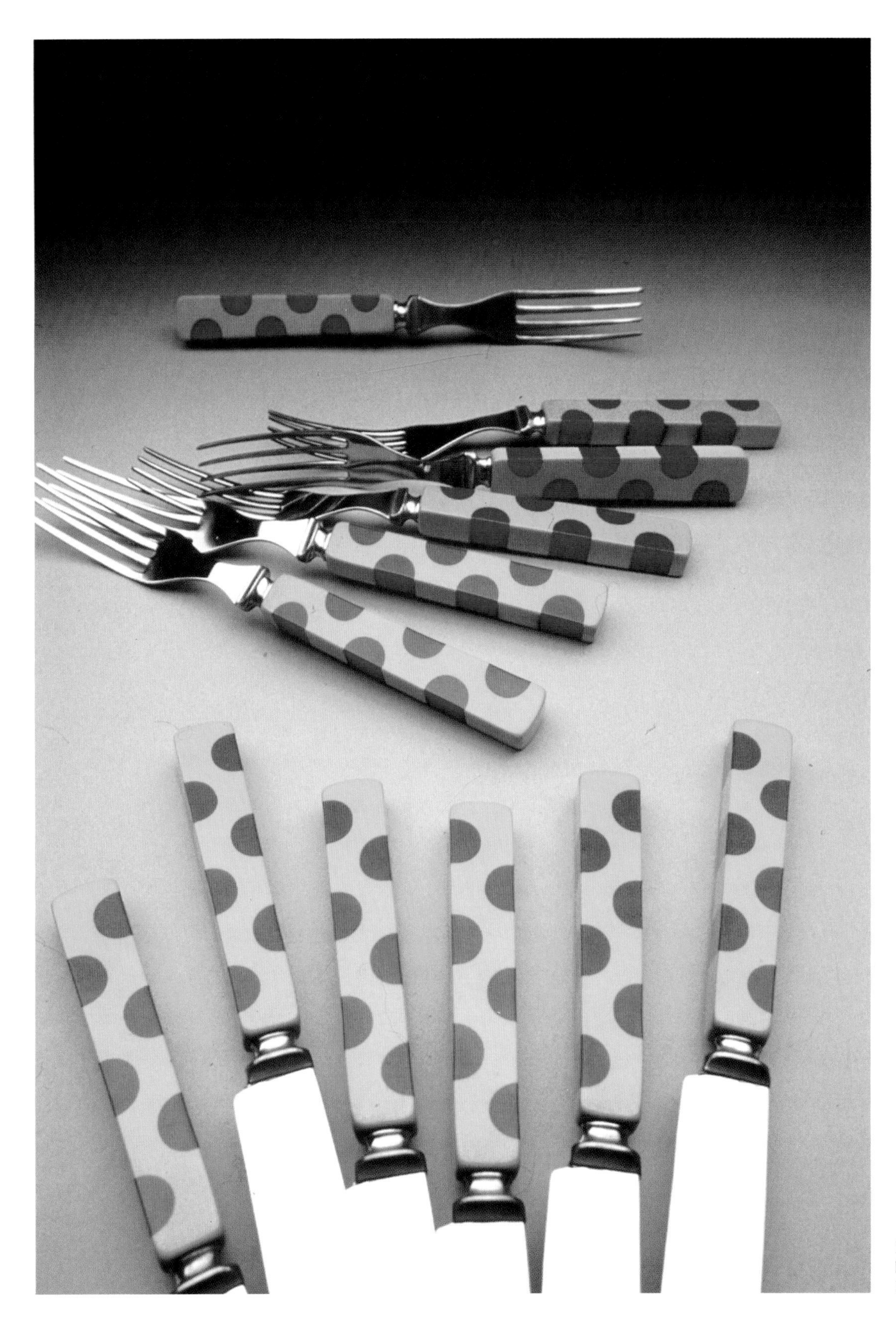

Decorative plastic-handled flatware. USA 1930s.

Napkin rings carved from cast phenolic resin. USA 1930s and 1940s.

and their close relationship with modernity. His main area of interest, as this excerpt from the chapter entitled 'Industrial Design and Commercial Art' demonstrates, is in the way in which the designer for industry can adapt his existing skills to designing plastic products. Gloag's pragmatic approach focuses on the way in which the physical requirements of the new materials make new demands on the designer which have to be thought out from scratch, and the established canons of 'good design' modified accordingly: He sees, for example, plastics offering new opportunities where both form and decoration are concerned. Finally he considers the innovative role in the areas of 'commercial art', particularly packaging and display design.

A firm commitment to plastics as symbols of progress, modernity and democracy characterise the first three contributions to this section, demonstrating why the 1930s and 1940s have been dubbed 'the Age of Plastics'. In her essay, entitled 'Perceptions of Plastics: A Study of Plastics in Britain 1945-1956', however, Claire Catterall looks back at the way in which, during this later period, that spirit of utopianism had been eclipsed, and how, in its place, there emerged a much more ambivalent attitude which suspected plastics of duplicity and vulgarity. This was, to some extent,

Bakelite Michelin man ashtray. 1930s.

associated with the strong anti-American feelings which existed in Britain at this time but it was also a direct result of the growing chasm that was visible between 'Establishment' and 'mass' culture. The protagonists of the former despised everything produced by the latter – Hollywood films, advertising, pulp novels etc. – and for them plastic products, produced and consumed en masse, were tarred with the same brush. This coincided

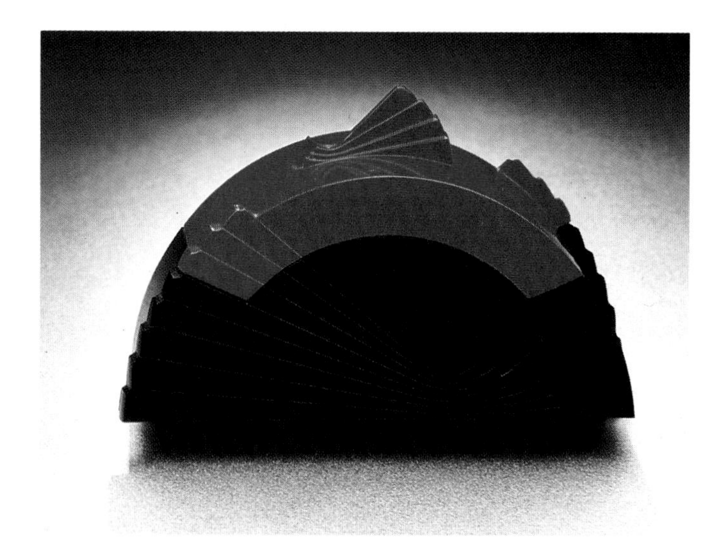

with the lowering of standards, in some quarters, of plastics manufacturing and, as a result, the materials' reputation sank to an all-time low.

While, on the whole, in the 1950s, Britain failed to sustain the cultural link between plastics and modernity that had been forged in the USA in earlier decades, Italy took a more pioneering and innovative approach towards the manufacture of plastic everyday goods in the post-war years. With the freshness of a country new to industrialisation Italy took, from the start, a positive approach towards the new materials and Italian manufacturing companies – among them Pirelli and Kartell – commissioned young, talented designers to work on the forms of such banal, everyday objects as buckets and washing-up bowls. For Italy plastics and 'quality' were synonymous and the concept of modernity was expressed through the most ordinary, mass-produced artefact. Through the late 1950s and 1960s the Italian plastics project developed into an extension of the pre-war Modern programme, manifested most clearly in a range of plastic furniture items designed by the leading architect-designers of the day. They served to breathe new life into the utopianism which had suffered such a blow in the immediate post-war years in Britain and the USA. G. Bosoni's article 'The Italian Way to Plastics' (reprinted from *Rassegna*, V. no. 14 2 June 1983, and translated from the Italian here by Marina Wallace) describes the Italian plastics project in some detail.

In spite of the pendulum swings in the cultural identity of plastics in the years 1918-60, the sense that the ideals of the Modern Movement could be realised by products made from the new materials kept returning to the surface and pointing the way forward. In the years after 1960, however, not only did this liaison become weaker but also the fundamentals of Modernism were themselves thrown into question. With the gradual emergence of a new design movement, loosely defined as 'Post-Modernism', however, plastics came back into the centre of the picture once again as the ideal materials to cope with the aesthetic and philosophical exigencies of the new spirit in design. Section Three will examine this new rapport between plastic products and the new cultural atmosphere.

Penny Sparke

'Cleopatra' manicure box in reinforced phenolic resin manufactured by the General Electric Company's Plastics Division. USA 1935.

Plastic products on market stalls. Italy 1960s.

PLASTICS IN THE AMERICAN MACHINE AGE, 1920-1950

Jeffrey L. Meikle

So much of the contemporary environment is moulded, fabricated, or constructed of plastics that we hardly notice its existence. But earlier, from about 1920 to 1950, the idea of a 'Plastic Age' conjured up visions of an infinite array of synthetic materials, products, and environments to be provided inexpensively, literally out of thin air, by wizards and miracle-workers of industrial chemistry. By 1940, most Americans had encountered synthetics in breakfast-cereal premiums, costume jewelry, radio cabinets, automobile steering wheels, nylon stockings, and Formica counter tops. None of these seemed exactly utopian, with the possible exception of nylon, which Du Pont unwisely claimed to be 'as strong as steel, as fine as the spider's web'.[1] Yet Americans did view plastics as miracle materials from which to shape the contours of a desired future. The story of how plastic gained and then lost its utopian significance illuminates the intersection of art, business, and technology during the American machine age.

Chorus girls from the Shepherd Show at the Princes Theatre, London, wearing nylon stockings imported from the USA. 1946.

Such early plastics as celluloid and Bakelite shared in a mystique generated by the chemical industry. In 1907 *Everybody's Magazine* informed its readers about the 'new synthetic chemistry' in an article on 'The Miracle-Workers: Modern Science in the Industrial World'. The author concluded that 'raw materials for building up' just about anything 'lie everywhere about us in abundance'. Scientists would never rest until they could make 'a loaf of bread or . . . a beefsteak' from 'a lump of coal, a glass of water, and a whiff of atmosphere'.[2] Other popular articles echoed this refrain, which became more pronounced after World War I – 'a war of chemists against chemists'. Edwin E. Slosson, a prolific populariser, reported on 'Chemistry in Everyday Life' in *The Mentor* in 1922. The human race, Slosson claimed, stood at the beginning of an era of material abundance made possible by coal – dirty, black coal – one of the most unappealing substances known. This limitless 'scrap heap of the vegetable

Advertisement for Bakelite Ltd. using designer John Vassos. *Modern Plastics*, USA July 1933.

kingdom' provided chemists with 'all sorts of useful materials' because coal contained in condensed form 'the quintessence of the forests of untold millenniums'. From sticky, evil-smelling coal tar, scientists synthesised not only dyes in 'all the colours of Joseph's coat' but also phenol, a major ingredient of Bakelite. Thanks to the plastics Bakelite and celluloid, the latter a wondrous 'chameleon material', imitations of amber, ebony, onyx, and alabaster were 'now within the reach of everyone'. The synthetic chemist thus acted as an 'agent of applied democracy' by making luxuries available as 'the common property of the masses'.[3]

John K. Mumford evoked an even more millennial tone in a promotional book, *The Story of Bakelite* (1924). At the time of the creation of life, nature began storing up 'waste heaps' of dead organic matter from which research chemists later derived 'colossal assets'. Bakelite seemed 'a wonder-stuff, the elements of which were prepared in the morning of the world, then laid away till civilization wanted it badly enough to hunt out its parts, . . . put them together and set them to work'. Mumford marvelled over Bakelite's 'Protean adaptability to many things', and the ease with which it could be moulded, but he found more miraculous the

Candlestick in catalin and metal. 1930s.

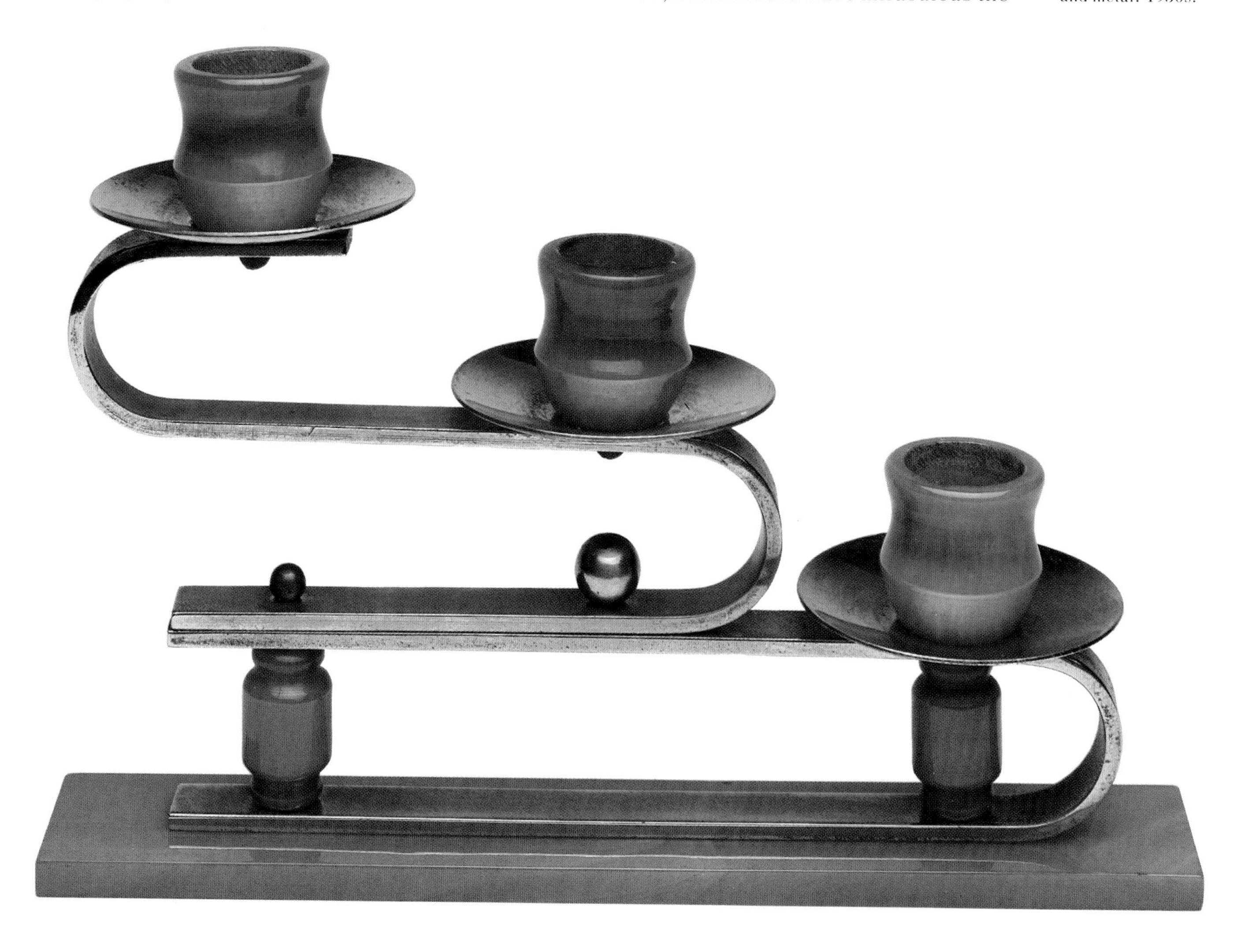

chemical reaction by which it set or hardened, after which it would 'continue to be "Bakelite" till kingdom come'. Out of the very stuff of death, Mumford suggested, came an indestructible material imbued with immortality. Chemists would indeed 'make a new world' by creating 'new substances . . . out of anything'.[4]

Popular magazine articles and books celebrated plastic during the 1920s and 1930s as a product of utopian magic, creating an artificial world of transcendent beauty and perfection from earth's commonest elements. Even *Fortune*, the nation's most intellectual business journal, entitled a 1936 review of the plastics industry with a biblically resonant phrase: 'What Man Has Joined Together . . .'. A more practical journal, *Business Week*, attributed the hoopla surrounding plastics to public ignorance and to 'the difficulty of translating into common terms the mysterious ways in which chemical processes move, their wonders to perform'. *Business Week*'s reporter savagely attacked 'popular stories of the "modern miracle" type, illustrated by figures symbolizing commercial research holding test tubes (usually by the wrong end)'. This sarcastic comment highlighted the degree to which the

Catalin radio manufactured by Fada. USA 1930s.

public did believe in modern miracles controlled by human agency.[5]

The view of plastics as miracle materials received extravagant expression in *Form and Re-form* (1930), written by designer Paul T. Frankl. 'Base materials', Frankle intoned, 'are transmuted into marvels of beauty' by industrial chemistry, which 'today rivals alchemy'. Plastics like Bakelite 'spoke' to modern man 'in the vernacular of the twentieth century' – 'the language of invention, of synthesis'. Unlike such pioneers as Parkes, Hyatt, and Baekeland, who intended their discoveries to imitate more expensive natural materials, Frankl asserted 'the autonomy of new media'. It was time to recognise the revolutionary nature of plastics and 'to create the grammar of these new materials'.[6] As a designer, Frankl was well aware that plastics were becoming more visible in design and architecture – areas where style, taste, and aesthetics received paramount attention. Plastics could express the beauty of the new machine age.

Several factors contributed to the success of plastics in consumer goods during the late 1920s and into the 1930s, such as increasing competition among materials suppliers, development of new plastics, and industrial design's new prominence. Competition became a factor in 1927 when the Bakelite patent on phenol-formaldehyde resin expired. Other manufacturers rushed to open phenolic plants, forcing down the price of Bakelite, and it became possible to use the material for inexpensive consumer products.[7] Competition also stimulated introduction of new phenolic plastics in a rainbow of colors. The Bakelite Corporation had normally supplied its material in black or dark brown to disguise fillers used in

Jewellery made of catalin and bakelite. USA. Circa 1935. **Collection: Susan Kelner Freeman**

molding compounds for strength. In 1927, taking advantage of the expired patent, the American Catalin Corporation offered 'Catalin, an insoluble, infusible cast phenolic resin of gem-like beauty and an unlimited color range, which in form of rods, tubes, sheets, or shapes can be machined on ordinary shop equipment'.[8] Used for toys, costume jewelry, chessmen, and decorative panels, Catalin required no fillers and therefore could be supplied in any solid, mottled, translucent, or transparent colour. Not to be outdone, Bakelite introduced its own cast resins.

The late 1920s also witnessed entirely new plastics to compete with celluloid and with phenolics. Cellulose acetate, developed by the Celluloid Corporation and marketed as Lumarith, provided a colourful, nonflammable celluloid substitute used for lighting fixtures and lamp shades.[9] An early vinyl introduced in 1928 under the name Vinylite was used for phonograph records, dentures, and beer-can linings. It gained exposure at the Chicago Century of Progress Exposition of 1933, where a 'house of the future' boasted a Vinylite interior-walls, floor, furniture, and many fixtures.[10] The most successful new plastics were urea formaldehyde resins developed to provide moulding compounds that, unlike Bakelite, offered an infinite colour range, including white, without sacrificing strength. The first urea formaldehyde plastic was Beetle, made available in the United States under British license in 1929.[11] Almost simultaneously an American firm, the Toledo Scale Company, funded research at the Mellon Institute in Pittsburgh to develop a urea plastic strong enough and light enough in colour to be used for the housing of a grocery-store scale.

Plaskon went on the market in 1931, but production of the scale, which required a forty-five-ton moulding press, the largest then in existence, did not begin until 1935.[12] Other major plastics available before the Second World War included clear, glass-like acrylics, marketed as Plexiglas by Rohm & Haas and as Lucite by Du Pont; and of course nylon, whose 'sheer magic' was demonstrated by young women at Du Pont's streamlined exhibit for the New York World's Fair.[13]

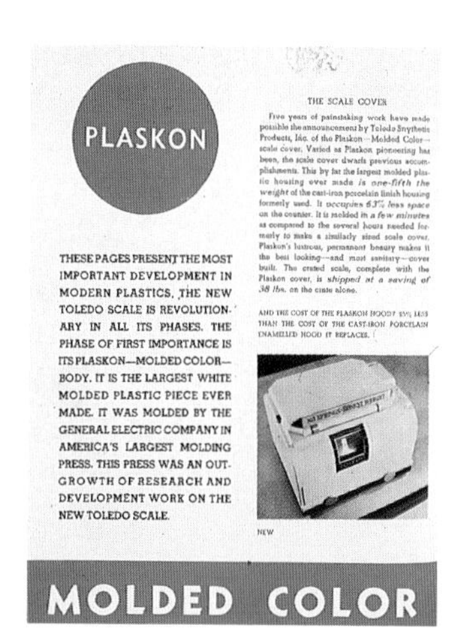

Far left: An advertisement for 'Beetle'. *Plastics and Moulded Products*. USA August 1930.

Left: Weighing scale made from 'Plaskon' urea, manufactured by the Toledo Scale Company. *Modern Plastics* USA 1934.

Popularity of plastics derived not only from introduction of new varieties. Strong economic motives dictated use of plastics in consumer products and ensured that designers employed them in futuristic machine-age styles. In a very real sense, the plastics industry and the new profession of industrial design developed together during the 1930s. Industrial designers like Walter Dorwin Teague, Henry Dreyfuss, Raymond Loewy, Norman Bel Geddes, and Harold Van Doren spent the decade redesigning consumer products to make them more attractive to potential purchasers. Businessmen hoped industrial design would solve the so-called underconsumption problem of the Depresssion. Dozens of products from toasters to refrigerators underwent face-lifting operations in an attempt to engineer economic recovery. If, as designers said, their profession was a 'depression baby', then the plastics industry was, in *Fortune*'s words, a 'child of the depression'.[14]

There was some truth to the claim that 'A Plastic a Day Keeps Depression Away'.[15] Designer Peter Muller-Munk recalled that 'plastics became almost the hallmark of "modern design" – as the mysterious and attractive solution for . . . any application requiring "eye appeal".'[16] There were very good reasons for this. To make products affordable, manufacturers had to reduce costs. As plastics became cheaper, they became attractive substitutes for traditional materials. In addition, products made from moulded plastics did not require expensive hand labour for assembling and finishing. A plastic radio case popped from its mould in a single piece, already coloured by a dye mixed into the moulding compound. Manufacturers hesitated only because of lingering fear that the public might reject plastic as an inferior substitute. The plastics industry mounted an aggressive campaign to suggest that synthetic materials had exactly the aesthetic qualities required for machine-age design. An executive of General Plastics summed up the campaign when he urged colleagues to 'make decoration symbolic of our modern age, using simple machine-cut forms to get that verve and dash which is so expressive of contemporary life'.[17]

The Bakelite Corporation led the way in convincing manufacturers to beautify products with plastic. In 1932 the firm held a symposium to acquaint designers with

An advertisement for Bakelite Ltd. using the designer, Raymond Loewy. *Modern Plastics*, USA June 1934.

An advertisement for Tenite depicting a desk phone, model 300, designed by Henry Dreyfuss. *Modern Plastics*, USA April 1936.

An advertisement for Plexiglas. *Modern Plastics*, USA October 1937.

Raymond Loewy says

..."The 'New Deal' in industrial design is establishing the triumph of beauty through simplicity"

THAT "manufacturers have discovered the buying public's inherent taste for simple beauty and its appreciation of fundamentally good lines" is the contention of Mr. Loewy. Of invaluable help to the designer in his search for simple, beautiful lines are Bakelite Materials.

The accompanying before and after illustrations of a duplicating machine provide a striking example of the improvement made possible through enlisting the aid of an artist-designer. "Before"—a mysterious appearing assembly of wheels, gears and cylinders, mounted on four iron legs, with a filing case precariously swung between them. "After"—a handsome cabinet of simple lines to serve as a pedestal, and a compact machine with operating mechanism concealed within attractive housings of lustrous black Bakelite Molded.

Product designers have discovered in Bakelite Materials, both Molded and Laminated, practical mediums for the interpretation of their ideas. The rich, lustrous finish of Bakelite Materials, their strength and permanence, the facility with which they may be formed, the many attractive colors available, make them invaluable in improving the design and increasing the saleability of many products.

By offering close and helpful cooperation with product designers, custom molders of Bakelite Materials will uncover many opportunities for increasing their volume of business. Write for a copy of new edition of 22M, "Bakelite Molded."

Mr. Loewy has served as body designer and consultant stylist to the Hupp Motor Car Corporation, as industrial designer and stylist for such institutions as Sears Roebuck, the Shelton Looms, Westinghouse Electric, Colonial Radio Corporation and others. The influence of Mr. Loewy's designs are reflected in a wide variety of industrial products.

Duplicating machine before and after being redesigned by Mr. Loewy. In the new design Bakelite Molded is used.

BAKELITE CORPORATION, 247 Park Avenue, New York, N.Y. . . . . 43 East Ohio Street, Chicago, Ill.
BAKELITE CORPORATION OF CANADA, LIMITED, 163 Dufferin Street, Toronto, Ontario, Canada

BAKELITE

THE MATERIAL OF A THOUSAND USES

June '34: X, 6 — Plastic Products — 205

Desk Monophone molded in colored TENITE by Automatic Electric Company

TENITE imparts lustrous color and exceptionally high strength to the new desk Monophone, which has broken sales records since its introduction. The limitless color range and superior shock resistance of Tenite make it the preferred plastic for many industrial and decorative uses. Write today for a new 52-page book on Tenite.

TENNESSEE EASTMAN CORPORATION *(Subsidiary of Eastman Kodak Co.)*, **KINGSPORT, TENN.**

APRIL 1936 — 39

Now being developed for molding—

PLEXIGLAS

THIS thermoplastic, the first solid acrylic resin to be manufactured on a commercial scale in the United States, has been on the market for over a year in the form of plane and curved sheets. Plexiglas* is characterized by the outstanding permanent properties of perfect colorless transparency, and stability against aging and weathering.

* Reg. U. S. Patent Office

CRYSTALITE

This granular acrylic thermoplastic molding material will be available in commercial quantities in the near future. The definite date of availability of Crystalite* will be announced in one of the next issues of "Modern Plastics."

Crystalite molded objects are characterized by colorless transparency, relatively high softening point, and freedom from warpage and shrinkage.

* Reg. U. S. Patent Office

RÖHM & HAAS COMPANY, INCORPORATED

222 West Washington Square PHILADELPHIA, PA.

"Plexiglas is on display at the Metals and Plastics Bureau, International Building, Rockefeller Center, New York City."

Phenolic fan, coffee grinder and 'Jumo' light. 1920s-1940s.

technical advantages and limitations of plastics as materials and also with their stylistic potential. Many businessmen already idolised the industrial designer as 'a wizard of gloss, the man with the airbrush who could take the manufacturer's widget, streamline its housing, add a bit of trim, and move it from twentieth to first place in its field'.[18] If industrial designers embraced plastics, then it seemed the battle was won. They needed little convincing. Over the next two years *Modern Plastics* and *Sales Management* ran a series of advertisements focusing on individual designers and their Bakelite products. Each ad featured a single product, each contained a small photograph and capsule biography touting the designer as celebrity, and each quoted the great man himself on the virtues of modern design. Bakelite applications ranged from knobs and handles to personal accessories like barometers and telephone indexes, from appliances like irons and washing machines to business equipment like mimeograph machines. Through it all ran the message that Bakelite – 'the material of a thousand uses' – would revitalise industry and reshape the environment.

The relationship between the plastics industry and design was symbiotic. As *Business Week* awkwardly phrased it in 1935, 'modernistic trends have greatly boosted the use of plastics in building, furniture and decoration, and contrariwise, plastics by their beauty have boosted modernism'.[19] As it turned out, the most economical methods of making moulds for plastic products lent themselves to machine-age styles. Harold Van Doren's 'Air-King' radio of 1933, one of the first with a plastic case, exhibited the zigzag setbacks and geometric ornament of Art Deco. As Van Doren pointed out, geometric designs – 'steps, ribs, circles, flutes, etc.' – could be cut into a mould by machine tools, eliminating the expense of hand labour. To design in a more intricate fashion would have made moulds too expensive.[20]. Even more appropriate to plastic mould technology was streamlining, the aerodynamic style of the 1930s. Low, sculptural, and flowing, streamlined design reflected American desire for frictionless flight into a future whose rounded forms would provide a protective, harmonious environment. As streamlining spread from locomotives and automobiles to stationary products like fans and vacuum cleaners, it received much criticism. But the public loved streamlining, and its defenders could argue that it made good sense for objects of moulded plastic. As late as 1946, nearly every engineer who discussed plastic product design emphasised the virtues of streamlining. A rounded or streamlined mould could be cut and polished by machine, but a mould with sharp edges and corners required expensive hand finishing. A rounded mould also permitted smooth flow of molten plastic to every area and surface. Copying wind-tunnel tests of aerodynamic engineers, plastics engineers even used dyes to trace the flow of material in variously shaped moulds and concluded that 'streamlined flow should be designed into both the inside and outside of a plastic part.[21] After a plastic radio case or electric razor was out of the mould, assembled, shipped to stores, and in the consumers' hands, rounded edges and corners provided protection from breakage. Rounded contours also brought out the reflective beauty of glossy plastic. Plastic and streamlining thus reinforced each other. One journalist even claimed that the requirements of plastic mould technology had inspired streamlining as a design style.[22] The statement was far fetched, but 'miracle materials' were indeed linked, through industrial design and styling, to the concept of an approaching technological utopia.

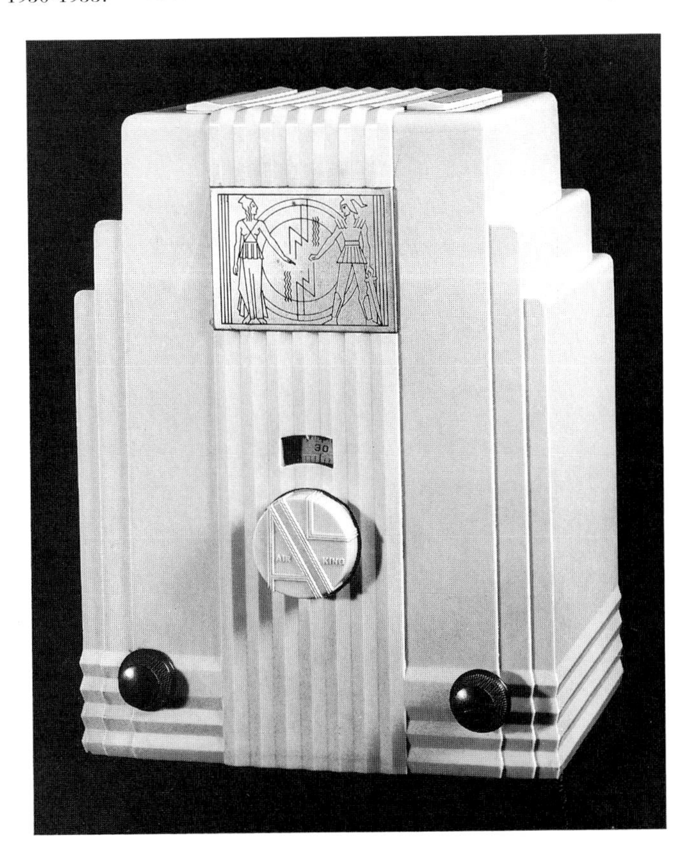

'Air-king' radio designed by Harold Van Doren and J.G. Rideout. USA 1930-1933.

Journalism reflected popular attitudes about new materials. In 1932, the Depression's worst year, *Literary Digest* ran an article on the 'Approach of the "Plastic Age"', while *Review of Reviews* tagged a summary of developments with a hopeful title, 'Synthetic Age . . . Era of Make-

Believe'. A few years later *Popular Science Monthly* returned to an earlier rhetoric when revealing how 'New Feats of Chemical Wizards Remake the World We Live In'. At the decade's end *National Geographic* described a fashion model 'clad from head to foot in artificial materials' as 'a startling symbol of this new artificial world risen so fast'. And *Popular Mechanics*, after observing that plastic was 'invading one field after another', claimed that 'the American of tomorrow . . . clothed in plastics from head to foot . . . will live in a plastics house, drive a plastics auto and fly in a plastics airplane'.[23]

Plastics utopianism received an eccentric boost from Henry Ford in November 1940. Then seventy-seven years old, Ford picked up an axe and swung it into the trunk lid of a custom-built 1941 automobile, while reporters took notes and photographers recorded the event. Rather than crumpling and losing its paint, the panel rebounded into shape and looked as good as new because it was made of tough plastic with embedded colour. Constructed of vegetable fibres (hemp, flax, and ramie) compressed in a phenolic resin, the panel seemed to promise realisation of Ford's dream that 'some day it would be possible to grow most of an automobile'. Eight months later, Ford unveiled a car with a body made entirely of fibre-phenolic panels and predicted that plastic cars would soon roll off the assembly line. But Ford's plastic car was never put to the test. As the United States approached entry into the Second World war, new plastic developments occurred secretly, and Americans were left with a vague notion that 'the technological novelty known as plastics' had 'graduated from its celluloid-and-Beetleware phase into an instrument of industrial revolution'.[24]

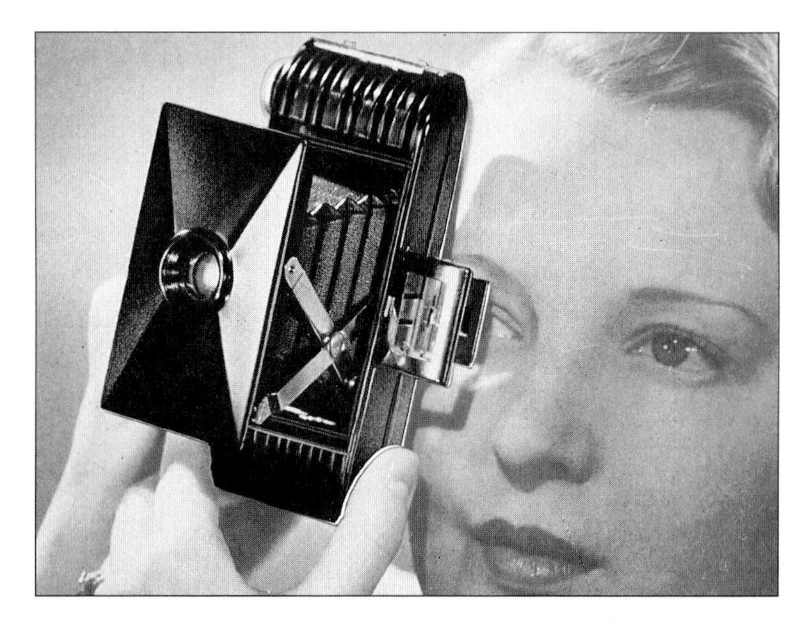

Walter Dorwin Teague. 'Bantam' camera manufactured by Kodak. 1936.

During the war, publicists continued to fuel utopian expectations about plastic. A little book called *Plastic Horizons* (1944), one of a series on 'Science for War and Peace', outlined the post-war potential of synthetics. The authors cited a typical wartime advertisement asserting that 'When the Minute Man returns to his Plow – it will have *Plastic* handles!'[25] The plastics industry expanded during the war to provide military equipment and also to replace materials no longer available for basic consumer goods. Soldiers encountered plastics in phenolic helmet liners and mortar fuses, vinyl raincoats, ethyl cellulose canteens, urea buttons, cellulose acetate bayonet scabbards and gas-mask parts, and melamine dishes, not to mention injection-moulded bugles.[26]The public was more interested in Plexiglas cockpit covers for aircraft, which suggested bubble-domed post-war automobiles, and in plastic-laminated gliders and light planes, which promised 'the family car of the air' or 'the Ford of the

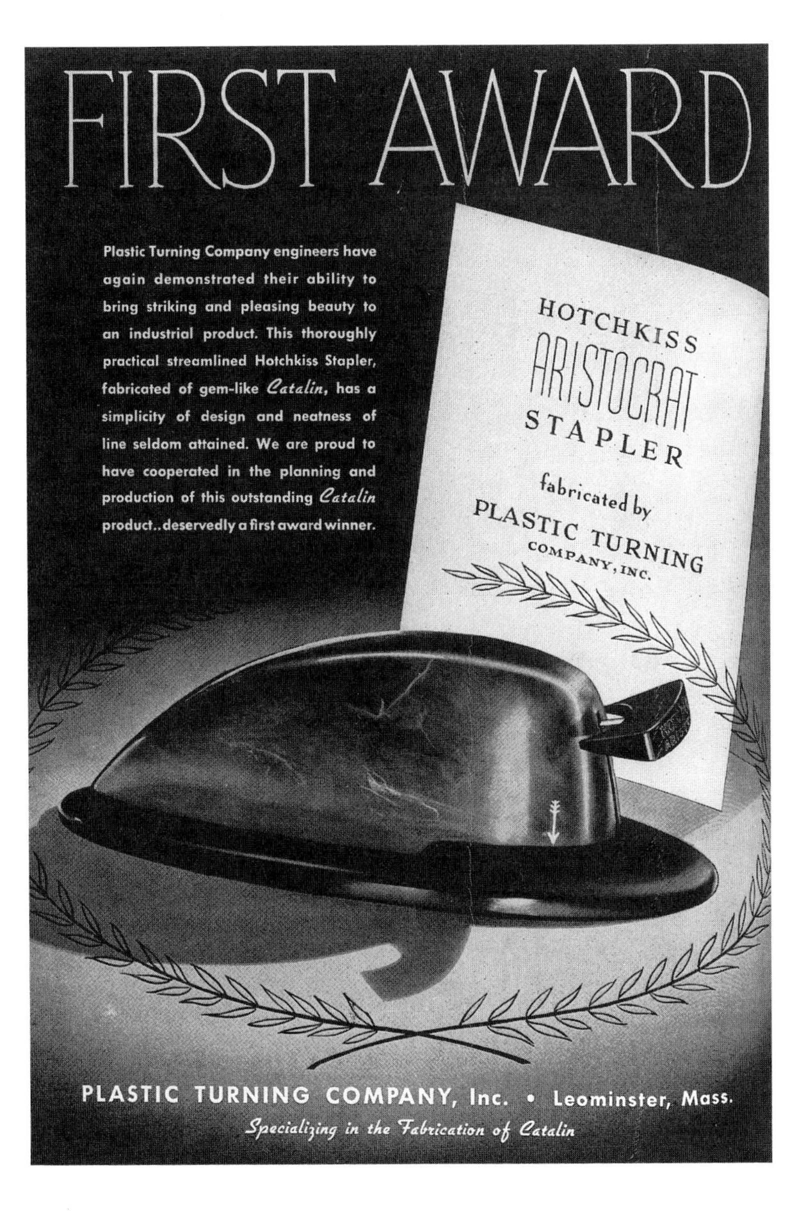

FIRST AWARD
Plastic Turning Company engineers have again demonstrated their ability to bring striking and pleasing beauty to an industrial product. This thoroughly practical streamlined Hotchkiss Stapler, fabricated of gem-like Catalin, has a simplicity of design and neatness of line seldom attained. We are proud to have cooperated in the planning and production of this outstanding Catalin product..deservedly a first award winner.
HOTCHKISS
ARISTOCRAT
STAPLER
fabricated by
PLASTIC TURNING
COMPANY, INC.
PLASTIC TURNING COMPANY, Inc. • Leominster, Mass.
Specializing in the Fabrication of Catalin

An advertisement for the 'Hotchkiss' stapler. *Modern Plastics*, USA November 1937.

skyways'.[27] Contributing to home-front morale, *Newsweek* announced that 'Test-tube Marvels of Wartime Promise a New Era in Plastics'. Combining the older vision of white-coated laboratory chemists and the newer vision of a post-war plastic cornucopia, *Newsweek* marvelled that 'molecule engineers' could actually 'draw blueprints of the kind of new molecules that they need for a given purpose'.[28] But plastics utopianism promised more than mere molecules – even synthetic ones – could deliver.

Plastics did compose an ever-greater percentage of materials used in post-war consumer goods, but its reputation plummeted. For one thing, cars and houses were *not* moulded of plastic, and the family airplane, of whatever material, never became a reality. More significant were some less obvious factors. During the war plastic had substituted for other materials – often in poorly conceived applications. According to one account, plastics were 'ersatz – something to be worried along with until more common materials are once more obtainable'.[29] After the war ended, people wanted 'genuine' materials. To make matters worse, many plastic products sold at war's end were made from poor-quality scrap. Consumers complained of combs dissolving in hair oil, dishes losing shape in hot water, and buttons becoming greasy blobs during dry cleaning.[30] When the industry tried to discipline its own members, changing design preferences undermined the effort.

Frankly synthetic materials, so popular during the 1930s, no longer appealed to Americans. As they confronted an uncertain post-war world, they turned away from machine-age styles to more traditional pseudo-historical styles reflecting nostalgia for a secure past. Plastic was indeed used more and more in furniture, fabrics, wall and floor coverings, decorative housewares, and appliances, but once again, as in the celluloid era, imitation took over. During the late 1940s, magazines like *Better Homes and Gardens* ran articles on 'how to put plastics together in a comfortable, normal house instead of a facsimile of a Statler cocktail bar'. Rather than being 'by nature shiny,

A selection of decorative pens cased in plastic. 1940s.

sleek, and a little too strange-looking for the living room', plastics could be 'homey, chintzy, and comfortable'. The goal was selecting plastic furnishings that 'don't look it'. The utopian fervor of plastics narrowed to one minor element. A housewife could clean synthetics, as everyone pointed out, with a mere wipe of a damp cloth.[31]

Whatever its usefulness to human life and comfort, plastics in the immediate post-war era inspired no vision greater than damp-cloth utopianism. Travelling full circle to its origin a hundred years earlier, plastic was employed for its imitative potential rather than as an expression of technological utopianism. In its dime-store manifestations, plastics had earned a reputation for being cheap and shoddy. It was a substitute, nothing more, nothing less, and fair game for any critic who considered American life superficial, phony, or abstracted from reality. This image problem has survived for decades across the social spectrum despite successive Euro, high-tech, retro, and Post-Modern appreciations of plastics. However, as more and more people cannot remember a world not pervaded by synthetic materials, plastics assume a neutral status, neither suffering the onus of imitation nor reflecting the promise of utopia. Plastics are being naturalised.

Jeffrey L. Meikle is Associate Professor of American Studies and Art History at the University of Texas at Austin. At present he is writing a cultural history of plastics in the United States.

Notes

1. C. M. A. Stine. Quoted in 'Du Pont Launches Synthetic Silk', *Business Week*, 29 Oct. 1938, p.18
2. H. Smith Williams, 'The Miracle-Workers: Modern Science in the Industrial World', in *Everybody's Magazine*, 17 (Oct. 1907), pp.497-8
3. E. A. Slosson. 'Chemistry in Everyday Life' in *The Mentor*, 10 (April 1922), pp.3-4, 7, 11-12
4. J. K. Mumford. *The Story of Bakelite*. New York: Robert L. Stillson, 1924, pp.7, 20, 22, 46, 51
5. 'What Man Has Joined Together . . .' in *Fortune*, 13 (March 1936), p.69; 'Plastics' Progress' in *Business Week*, 21 Dec. 1935, p.17
6. P. T. Frankl. *Form and Re-Form: A Practical handbook of Modern Interiors*. New York: Harper, 1930, p.163
7. See Williams Haynes. *American Chemical Industry: The Merger Era*. New York: Van Nostrand, 1948, p.348
8. Quoted by Haynes, op. cit., note 7, p.349
9. 'Plastics' Progress', loc. cit., note 5, p.16; Haynes, op. cit., note 7, pp.350-1; and Williams Haynes, *American Chemical Industry: Decade of New Products*. New York: Van Nostrand, 1954, pp.330-331
10. A. E. Buchanan, Jr. 'Synthetic Houses' in *Scientific American*, 149 (Oct. 1933), p.149; J. Singer. *Plastics in Building*, London Architectural Press, 1952, pp.25-7; and Haynes, *Decade of New Products*, pp.338-9
11. Haynes, op. cit., note 7, p.354; Haynes, op. cit. note 9 *Decade of New Products*, pp.335-6
12. A. M. Howald. 'Systematic Study Develops New Resin Molding Compound' in *Chemical and Metallurgical Engineering*, 38 (Oct. 1931), pp.583-4; J. L. Rogers. 'Plaskon, a New Molding Compound the Result of Planned Research', in *Plastics and Molded Products*, 7 (Dec. 1931), pp.664-5, 687; 'Giant Plastic Moulding Press Produces Large Weighing Scale Housings' in *Iron Age*, 136 (29 Aug. 1935), pp.13-14; and H. D. Bennett, 'Pushing Back Frontiers', in *Modern Plastics*, 13 (Sept. 1935), pp.25-7, 30-2
13. F. D. Morris. 'Sheer Magic', *Collier's*, 105 (13 April 1940), pp.13, 69-71
14. N. G. Horwitt. 'Plans for Tomorrow: A Seminar in Creative Design' in *Advertising Arts* July 1934, p.29; 'What Man Has Joined Together . . .' op. cit., note 5 p.69
15. 'A Plastic a Day Keeps Depression Away' in *Chemical and Metallurgical Engineering*, 40 (May 1933), p.248
16. P. Muller-Munk. 'The Future of Product Design' in *Modern Plastics*, (20 June 1943), pp.77, 144
17. F. E. Brill. 'Some Hints on Molded Design' in *Plastic Products*, 9 (April 1933), p.55
18. P. Muller-Munk. Quoted by Seymour Freedgood. 'Odd Business, This Industrial Design' in *Fortune*, 59 (Feb. 1959) p.132
19. 'New Jobs for Plastics' in *Business Week*, 28 Dec. 1935, p.17
20. H. Van Doren. 'A Designer Speaks His Mind' in *Modern Plastics*, Sept. 1934, p.24
21. 'Small Radios—Today and Tomorrow' in *Modern Plastics*, 17 (March 1940), p.80; R. L. Davis and R. D. Beck. *Applied Plastic Product Design*. New York: Prentice-Hall, 1946, p.44
22. R. P. Calt. 'A New Design for Industry' in *Atlantic Monthly*, 164 (Oct. 1939), pp.541-2
23. *Literary Digest*, 112 (2 Jan. 1932), p.42; *Review of Reviews*, 86 (Nov. 1932), pp.62-3; Alden P. Armagnac, 'New Feats' in *Popular Science Monthly*, 129 (July 1936), pp.9-11, 109; F. Simpich, 'Chemists Make a New World' in *National Geographic Magazine*, 76 (Nov. 1939), p.601; and J. P. Leggett 'The Era of Plastics' in *Popular Mechanics Magazine*, 73 (May 1940), pp. 130A, 658
24. 'Plastic Fords' in *Time*, 36 (11 Nov. 1940), p.65; 'Ford from the Farm' in *Newsweek*, 18 (25 Aug. 1941), p.39; and 'Plastic Ford Unveiled' in *Time*, 38 (25 Aug. 1941), p.63. On Ford's involvement with agricultural plastics see R. Wik. 'Henry Ford's Science and Technology for Rural America' in *Technology and Culture*, 3 (Summer 1962), pp.247-8; D. L. Lewis. 'Henry Ford's Plastic Car' in *Michigan History*, 56 (Winter 1972), pp.319-30
25. B. H. Weil and V. J. Anhorn. *Plastic Horizons*. Lancaster, Pa. Jacques Cattell Press, 1944, p.130
26. 'In the News' in *Modern Plastics*, 20 (May 1943), p.114
27. W. Ward Jackson. 'The Future of Plastics in Aviation' in *Modern Plastics*, 21 (Jan. 1944), P.176; Forrest Davis. 'Airplanes, Unlimited!' in *Scientific American*, 161 (July 1939), p.17
28. 'Test-Tube Marvels' in *Newsweek*, 21 (17 May, 1943), p.42
29. L. H. Woodman. 'Miracles? . . . Maybe' in *Scientific Monthly*, 58 (June 1944), p.421
30. Examples are from J. J. Pyle. 'New Horizons in Plastics' in *Science Digest*, 18 (Aug. 1945), pp.85-6; 'The Buyer Is Reaching for His Crown' in *Modern Plastics*, 24 (Feb. 1947), p.5; and 'A 1950 Guide to the Plastics' in *Fortune*, 41 (May 1950), p.111
31. Quotations are from 'Plastics: A Way to a Better More Carefree Life' in *House Beautiful*, 89 (Oct. 1947), pp.123, 141; C. Holbrook and W. Adams. 'Dogs, Kids, Husbands: How to Furnish a House So They Can't Hurt It' in *Better Homes and Gardens*, 27 (March 1949), pp.37-9

'Radio Nurse', in bakelite, designed by Isamu Noguchi for the Zenith Radio Corporation. USA 1938.

PLASTICS AND THE FUTURE

V. E. Yarsley and E. G. Couzens

SOME DAY YOU'LL BATHE IN A MOLDED TUB
BUT NOT YET, BROTHER, NOT YET!
Altho new materials and new applications of old materials are falling on the Plastics Industry, like rain on an April morning, not all of them are practical. Sometimes Cost is the Simon Legree. Sometimes the characteristics of the available materials set the limit.
When the enthusiasm of the raw material manufacturer carries him off the deep end, it's the molder who must outline the possibilities of trouble because his responsibility is that of a Consultant—he must see that the Future is coaxed into the Present reasonably and be sure that the pants fit the growing child.
We at Boonton pride ourselves on knowing the available materials and having all the available equipment. If bath tubs are not practical today we'll tell you why and maybe how soon, but we won't mold them for you regardless just because it's a fat job and you have plenty of money, unless you know the gamble.
Our existence depends on sound, honest, practical advice based on a hundred years of collective experience—yours for the asking.
Our slogan—the latch string is always out—is literally true. Our plant and organization is to be considered as your private molding department. We don't want you to get your neck in a noose because it's our neck, too, nor do we want you to miss any good bets for the same reason.
"A Ready Reference for Plastics." Written for the layman, this unique handbook explains the uses and characteristics of plastics in plain, non-technical language. Write for FREE copy.
PLASTICS
BOONTON MOLDING COMPANY
MOLDERS OF PLASTICS · PHENOLICS · UREAS · CELLULOSE · ACETATES
BOONTON · NEW JERSEY · Tel. Boonton 8-0991
N. Y. Office—30 Church St.—COrtlandt 7-7971
10 MODERN PLASTICS

An advertisement for a moulded plastic baby bath. *Modern Plastics*, February 1938.

Let us try to imagine a dweller in the 'Plastic Age' that is already upon us. This creature of our imagination, this 'Plastic Man', will come into a world of colour and bright shining surfaces, where childish hands find nothing to break, no sharp edges or corners to cut or graze, no crevices to harbour dirt or germs, because, being a child, his parents will see to it that he is surrounded on every side by this tough, safe, clean material which human thought has created. The walls of his nursery, all the articles of his toilet, his bath and certain other necessities of his small life, all his toys, his cot, the moulded light perambulator in which he takes the air, the teething ring he bites, the unbreakable bottle he feeds from, later all the equipment for his more mature daily meals, the trays, the spoons and mugs, all will be plastic, brightly self-coloured and patterned with every design likely to please his childish mind. As he grows up he cleans his teeth and brushes his hair with plastic bristles, clothes himself in plastic clothes of synthetic silk and wool fastened with plastic zip-fasteners, wears shoes of plastic and textiles covered with a plastic finish, writes his lessons with a plastic pen and does his lessons with books bound in plastic. He sits in a new kind of schoolroom with shining unscuffable walls at a moulded desk, warm and smooth and clean to the touch, unsplinterable, without angles or projections.

Baby's rattle made of bakelite. 1940s.

The windows of this school, curtained with plastic-faced cloth entirely grease- and dirt-proof, are unbreakable, and transmit the life giving ultra-violet rays, and the frames, like those of his house, are of moulded plastic, light and easy to open, and never requiring any paint to prevent them from warping or rusting. Like his home, too, the plastic floors are silent and dustless, but lest this picture seem too coldly hygienic, remember that everywhere there is a riot of colour and every kind of surface from dull matt to a mirror-finish that circumstances demand.

The very blackboard, now a pleasant shade of dark green, is of unscratchable matt plastic, and if corporal punishment has survived into this plastic age, he may have the privilege of being beaten with a synthetic plastic cane, far more tough and resilient than the one which Nature, unaided, can provide.

Back in his home he still finds the universal plastic environment. Once again the walls of his rooms are built with panels and plastic doors covered with beautiful veneers, either real or synthetic, protected by a transparent plastic finish. The bathroom will be all plastic: no ceramic tiles now to crack or flake or enamel bath to do likewise, but all made of heat-resisting tough plastic, white or pastel shade, with plastic taps and even plastic pipes which can be cemented together in a few moments without the clumsy apparatus of the plumber's blowlamp and solder. In all the living-rooms and bedrooms every kind of furniture is built up of moulded plastic sections, the wood formerly used for structure now being used in conjunction with plastic for purely decorative purposes. Even the armchairs and divans will be plastic, constructed of woven plastic 'reeds' and strips in every conceivable colour, warm, pliable and clean. Every small article with which man provides himself for his necessity or luxury and which does not require metal for a cutting edge, or for electrical purposes, or for heat resistance, will be plastic, and

Child's bandalasta tea-set.
Great Britain 1930s.

where metal parts are needed, as in a razor, safety or electric, or a sewing machine or an electrical device, all the casings and solid structures which carry the metal parts will be moulded plastic.

For beauty also, the 'Plastic Man' will turn to plastics. Lampshades and stands, screens, electroliers, all will be available in beautiful transparent glass-like materials in every imaginable form; carved and engraved transparent plastics, illuminated by concealed lighting, bouquets of flowers miraculously preserved in geometrical masses of brilliant, clear, transparent plastic, bowls and vases of every shape and colour and more transparent than glass, will all be accessible to the well to do, while for everyone the creative instinct can find satisfaction in the making of such objects from the readily accessible and inexpensive plastic masses which they will be taught to fashion in the craft schools of the future.

Better material for better light
JULY 1936 23

An advertisement for plastic lamp-shades. *Modern Plastics*, USA July 1936.

Outside the home, the same universal rule of plastics holds. Even the tennis-racquets, golf-clubs and fishing-tackle which employ the leisure of the 'Plastic Man' will be entirely made of plastic, and when he comes to travel he will find it everywhere. The motor-car, already employing many pounds of plastic in its construction, will then be almost wholly plastic in its externals, probably only the engine, transmission, chassis and wheels being made of metal. As for the aeroplane, the engine alone will provide an outlet for metal, for here the astonishing lightness of plastics will revolutionize construction. The wings, struts, fuselage, cabin, petrol tanks and all interior fittings will be mass-produced from reinforced plastic, making the standardized aeroplane the motor-car of the future. In buses, trains and ships, the same phenomenon will be observed, more and more actual structure being taken over by plastics as the possibilities of its low density-strength relation are better understood. Ships, air-conditioned and insulated with plastic insulation made from film, may even be finished externally with non-inflammable water-resisting plates mounted on the steel structure, and the 'Plastic Man' will undoubtedly go boating and yachting in small vessels built of moulded plastic sections.

In industry it will be the same story. In the office, where standardization is more acceptable than in the home, the whole interior surfacing and all the fittings except, perhaps parts of the typewriter, will be plastic. The trays, the filing cabinets, the loose-leaf books, the telephone, all the desk

'Coronet' camera, with bakelite case. USA 1930s.

equipment, the very desks themselves will be made of this universal material, because of its durability, surface attractiveness and eminent suitability for clean smooth design. In the factory, of necessity, plastics will play a more ancillary part, but every kind of housing or casing, all constructional members where lightness is required, and all repetition parts such as those that are required in great industries like spinning and weaving and which do not have to stand undue stresses, will be made of plastics. In the electrical industry, except for conductors and magnets for which metals remain essential, almost everything will be plastic; from massive high-tension insulators taller than a man to tiny short-wave radio parts, the coverings of all cables and flexes, the insulating wrappings of conductors in motors and alternators, the dielectric inserts in condensers will all find their material of construction in a suitable plastic. In fact, the manufacturer of the future will say, not 'of what material shall I make this article?' but 'what kind of plastic shall I use?' and the function of metal machines will become more and more that of making plastic articles.

But now our 'Plastic Man' is getting tired and old. His own teeth are gone and he wears a plastic denture with 'silent' plastic teeth and spectacles of plastic with plastic lenses, and combs his scant hairs with a plastic comb. He still takes photographs on plastic films with a camera moulded from plastic with a plastic lens, listens to the wireless through a set encased in plastic, with plastic insulation, dial and tuning buttons, sits in the cinema in plastic seats watching a picture projected from a plastic film, or stays at home playing with plastic playing cards and moulded chessmen on a plastic board, until at last he sinks into his grave hygienically enclosed in a plastic coffin.

'Plastic' coffin or not, this was the inevitable end, but in how much brighter and cleaner a world he has lived than that which preceded the plastics age. It is a world free from moth and rust and full of colour, a world largely built up of synthetic materials made from the most universally distributed substances, a world in which nations are more and more independent of localised natural resources, a world in which man, like a magician, makes what he wants for almost every need, out of what is beneath him and around him, coal, water and air. When the dust and smoke of the present conflict have blown away and rebuilding has well begun, science will return with new powers and resources to its proper creative task. Then we shall see growing up around us a new, brighter, cleaner and more beautiful world, an environment not subject to the haphazard distribution of nations' resources but built to order, the perfect expression of the new spirit of planned scientific control, the Plastics Age.

Interior for the Pierce Foundation making extensive use of plastic materials. USA 1933.

Left: Dentures, toothbrushes and mugs, all in Vinylite. *Modern Plastics*, USA. 1933.

Right: Multi-coloured mouldings in 'Vinylite' resin. USA 1933.

INDUSTRIAL DESIGN AND COMMERCIAL ART

John Gloag

The reaction of the industrial designer to the manifold gifts of the plastics family, is first to discover how stresses and strains, tensions and weights are affected by the materials; what limitations on shape and size are imposed by the various fabricating methods; whether surfaces are resistant to wear and tear and what upkeep, if any, they require. Equipped with such knowledge, he tackles with an open mind the problem of designing, say, a chair. He only has to accommodate the contours of the human frame, and although the posture adopted by most Europeans when they are seated has changed slightly during the last century, we are still vertebrates, even though we no longer care to sit bolt upright, as our great-grandfathers did, and lounge a few inches nearer the floor than they would have thought consistent either with dignity or decency. Since the early sixteenth century, chairs in England have been supported by legs; before that they were boxes with a high back and solid arms rising above the top of the box. Woodworkers learned how to economise in their material, and to trust its strength in new ways; so four legs, linked and braced by an underframe, eventually supported the chair seat. Then the underframe was eliminated, and the legs were tapered. For a couple of centuries a progressive refinement of design gave increasing elegance to the chair; then the slimming process stopped, and throughout the Victorian period the chair grew bloated and sank down to the floor; its legs bulged and thickened, its swollen feet were shod with castors. Now, the industrial designer has the problem of making a comfortable seat with 'limitless control of material'. The result may be a two-piece chair: a back and a seat in one piece, and a curved underframe in one piece to support it. That is, perhaps, an excessively simple solution; but it is not an impossible one, and I have mentioned it to illustrate the liberation of design from ancient limitations and pre-conceived notions about the form of anything that occurs when an industrial designer is working out a problem with plastics.

The second division of industrial design is complicated by the existence in this country and elsewhere of a vast number of so-called specialists in decoration, These enthusiastic people, many of them without taste of training, grasp at every opportunity for ornament, and once they realise the prolific capacity of plastics for enlivening surfaces and making blobs and knobs and wriggling shapes, they will insist on giving us all the colours of the rainbow plus all the confusion of 'the morning after'. But there are many able designers whose gifts can be productively engaged in the study of plastics in connection with industrial decorative art. In the devising of patterns for the surface decoration of some types of plastics their judgement and skill would be invaluable.

Natural and pearlised PVC belts. England 1940s.

Plastic tie-press. 1930s.

The trained designer will know how to handle such opportunities; though the judgment of the untrained designer may, like some mediaeval English monarchs, die of a surfeit. For interior decoration, for lighting fittings and ornamental articles, bowls, vases and lampshades, plastics have already become established in use. They were used extensively, for example, on the Cunard-White-Star liner, *Queen Mary*. Apart from interior decoration, the future of plastics in the fashion world carries us beyond the consideration of industrial design. To the designers of such things as costume jewellery, fancy buttons, buckles, handbags, compacts, shoes, belts – all the appurtenances of feminine toilet – plastics bring a new treasury of ideas. Again, the work of textile designers must in time be profoundly affected by the increasing use of specialised types of plastic fabrics. Like the progressive potters, textile manufacturers have worked out many fruitful partnerships with designers, and the study of plastics must inevitably initiate fresh design research work in their industry.

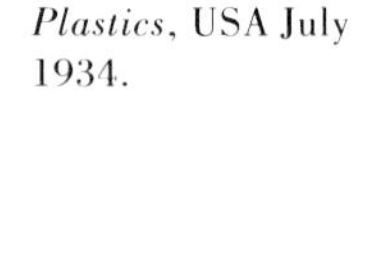

An advertisement for the American Catalin Corporation. *Modern Plastics*, USA July 1934.

For instance, there is a type of plastic which is built up from layers of paper, textiles or even thin veneers of wood. The layers are impregnated and bound together with a phenolic resin. These laminated plastics may be plain in colour, with smooth, hard, glass-like polished surfaces; or a satin finish or texture may be imparted to the surface from patterns stamped on the metal sheets of the press wherein the sheets are formed. Patterned paper or textiles may be used for the top layer of the lamination, or the plain surface may be ornamented by cutting out inlays through the sheets before compression.

The mastery of decorative industrial art attained by some of our artists and designers in ceramics and glass, suggests the richness and variety of talent that could bestow

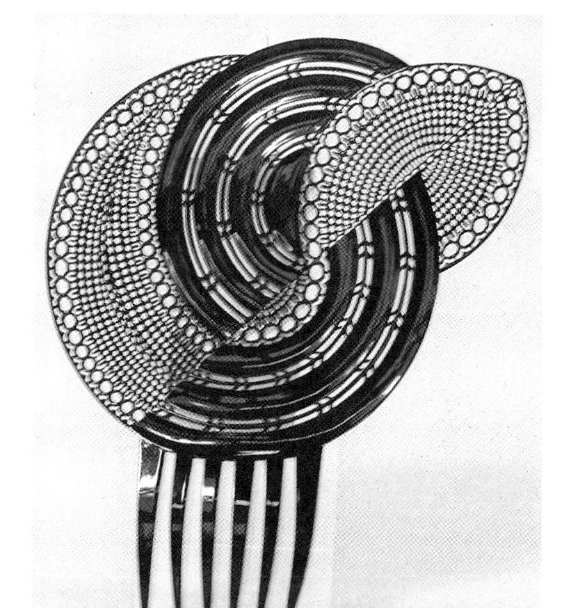

'Art Deco' comb. France, 1930s.

Lady's hat in cellulose acetate. *Modern Plastics*, USA July 1938.

comparable distinction on things produced in the new 'fifth class' of materials. This mastery is exemplified by the work of artists like the late Eric Ravilious for Wedgwood pottery; by Keith Murray's work for the same firm, and his designs in domestic glass for Stevens and Williams Limited; by the decorative treatments carried out for Pilkington Brothers Limited on various forms of glass, by Kenneth Cheesman, Sigmund Politzer and Hector Whistler, and by the patterns for glass designed by Paul Nash and R.A. Duncan for Chance Brothers and Company Limited. Only a few names have been mentioned of men who excel in this field: there are many others.

Having briefly examined the two divisions of industrial design in relation to plastics, commercial art may now be considered. Commercial art is not concerned with the form, function or production of goods, but only with their presentation to consumers. It has, perhaps, greater vivacity and certainly less permanence than industrial design. The term covers all those branches of artistic activity which assist the distribution of goods, such as press advertising, posters, booklets and leaflets, the labelling, packaging and display of merchandise, the design of display material in shops and stores, and the design of exhibition stands. It commands the services of commercial artists, typographers, packaging and display specialists; and to some of these people plastics will furnish opportunities for original experiment. It is obvious that in the packaging and display of goods, there will be an increasing use of plastics, particularly of transparent and translucent varieties.

Before 1939 there had been a great increase in the use of plastics for packs, for stoppers on bottles and tins, and for wrapping. But added gaiety and dramatically luxurious effects for the packing of such things as toilet preparations only represent one aspect of the service that can be rendered by plastics. Here is an example of the way the form and character of a container could be changed by an imaginative designer. The problem is presented by very small tablets of a drug, usually contained in a bottle which can be slipped into the waistcoat pocket; but the minuteness of the bottle causes a difficulty, for it only allows two restricted surfaces for accommodating a label, and on that label it is essential to set forth explicit directions for taking the tablets, their ingredients, and the name of the makers. This information has to appear in four languages, and even the use of a label that encircles the bottle cannot solve the typographical problem satisfactorily; so perforce it is solved unsatisfactorily, and the smallest type that can be set by a compositor is reduced still further by photography, until the essential material is crammed, illegibly, into the space available. The use of a transparent plastic would allow a packaging designer to produce a disc pack, smaller in diameter than a watch case and not much thicker than, say, three half-crown pieces. This disc pack would have a simple, screw-thread closure; the two sides of the disc would offer a much larger area than the bottle for printing the directions and so forth, and possibly the material could be printed direct upon the plastic surface.

A convenient form of pack for tablets has been in use for some years, consisting of sheets of transparent material, enclosing the tablets in parallel rows, so one or more may be torn off easily. Creative thinking by commercial artists and packaging designers

Bear book-ends made of catalin. 1930s.

may change the manner in which all kinds of goods are delivered to consumers, and shopping in the future may not only be more exciting, but far more convenient.

The growing influence of plastics in the display of goods was demonstrated by the New York World Fair in 1939. I observed, both in New York and in the San Francisco World Fair, which I visited in that year, a fresh approach to display problems which suggested that some remarkable ideas about the alliance of light with transparent and translucent materials were encouraging designers of display to make experiments. The enclosing of objects in irregular masses of transparent plastic, so that they interrupted and redistributed a beam of tinted light; the bubble delicacy of barely discernible display units, allowed goods to be arranged on almost invisible shelves and supports; an infinity of variations for edge lighting, with artificial light transmitted from edge to edge of transparent sheets of polystyrene – all these and many other partnerships between plastics, glass, metal and light were then being explored by American designers. The significance of such work had not of course escaped British designers; but the war interrupted developments.

It is the purpose of this book briefly to examine plastics in relation to industrial design: to attempt even a casual review of the ramifications of commercial art is beyond its scope. The subject has only been touched on in this chapter to avoid any possibility of the functions of industrial designer and commercial artist becoming confused. I do not imply that they are rigidly segregated: technical skill illumined by imagination will always transcend the tidy docketing of official registers and academic institutions. By virtue of training and creative ability the same man may be an industrial designer and a commercial artist; obviously no label, used merely for convenience, can affect his gifts or knowledge when he is solving a problem; but if the problem he solves is concerned with the shape, function and finish of goods, then he is practising industrial design; if his talents are exerted for devising methods of packing or exhibiting those goods, then he is practising commercial art.

If our designers are allowed to do their best work in collaboration with manufacturers who are making goods wholly or partly from plastics, the public at home will certainly be confronted with the results of genuine imagination, and we shall woo markets abroad with the help of a vigorous salesman whose salary never appears on the pay roll of a sales department.

Colour Perfection
RHC
STYRON
POLYSTYRENE
GRANULES
A DOW PRODUCT
The Right Plastic
for Injection,
Compression
and Extrusion
Versatile
Colourful
Economical
Serviceable
Stable
"FREE FROM DUST
AS RAIN-WASHED
AIR"
R.H.COLE & COMPANY LTD.
'Phone WHITEHALL 0711-2-3
'Grams, GERATOLE, PHONE, LONDON.
2 CAXTON STREET, WESTMINSTER, S.W.I.

PERCEPTIONS OF PLASTICS: A STUDY OF PLASTICS IN BRITAIN, 1945-1956

Claire Catterall

The plastics industry emerged from the Second World War with reason to be optimistic for its future. With the heroic incorporation of polythene into the radar system and successful developments in the use of PVC and polystyrene,[1] plastics had been given the chance to prove their potential and had at last tasted victory over their pre-war taboo of being thought 'merely substitute materials'. Armed with its revolutionary new recruits and spurred on by thoughts of the seemingly limitless potential for peace-time applications, the plastics industry looked forward to what appeared to be the dawn of a new era for its product.

Britain in the post-war years, however, did not welcome plastics as the industry had hoped it would. Plastics suffered a disastrous initial introduction into the public consciousness, which left the consumer thinking them only cheap-looking, unreliable and short-lived.[2]

Part of the problem lay with responses to the newly emergent concept of 'mass culture'. The British Establishment was suspect of it and the artefacts it engendered – advertisements, Hollywood films, streamlined cars, pulp novels among them.

An advertisement for Styron showing a range of objects, from the pre-war period, made of polystyrene. *British Plastics.*

Underlying this hatred of mass culture and acting as a focus point around which much of the fear and criticism could be levelled was what Dick Hebdige has described as the 'spectre of Americanisation'.[3] Resentment of America was built upon Britain's military and economic dependence, a fact which was exacerbated after the war by the continued reliance on American military presence and by Britain's declining status as a world power coupled with the simultaneous rise in America's international prestige. The British Press played up this hostility, often focusing attention on American GIs who were seen to be a subversive and unsettling influence. American automation processes and consumer goods, or objects designed and manufactured on American lines were also singled out for criticism.

Plastics were seen to be part and parcel of the American phenomenon. There was a visible manifestation of mass-production and rapid automation processes; their bright colours and shiny surfaces were easily equated with American brashness; their very 'syntheticness' was proof of their lack of validity in cultural terms. Plastics were the usurper, like the American, of all that was held sacred in the British consensus of tradition and value. Writers such as Evelyn Waugh can be seen to be defending traditional social privilege in their works through the delineation of a previously submerged set of values and assumptions.

This is most starkly seen in his exchange with Nancy Mitford over 'U' and 'Non-U', a system of codification designed to clarify and redefine social boundaries. In his last book '*The Ordeal of Gilbert Pinfold* (1957), Waugh continues to show his distress at the planing down of social and cultural distinctions.

'. . . his strongest tastes were negative. He abhorred plastics, Picasso, sun-bathing and jazz – everything in fact that had happened in his own lifetime. The tiny kindling of charity which came to him through his religion sufficed only to temper his disgust and change it to boredom . . . He wished no-one ill, but he looked at the world "sub specie neternitatus" and he found it as flat as a map; except when, rather often, personal arrogance intruded. Then he would come tumbling from his exalted point of observation. Shocked by a bottle of wine, an impertinent stranger or a fault in syntax.'[4]

Pinfold testifies to the wish to maintain class differences that was clearly prevalent in post-war Britain. For our purposes, however, it is Waugh's inclusion of plastics in his list which is most illuminating. Plastics were clearly equated with the horrors of modern culture, and by listing them with other themes of social and cultural taboo he was linking them with suggestions of subversion, crime, sex, cosmetics, increased leisure and inertia, social ignorance and faux pas.

This attitude was not confined to writers such as Waugh but could be found across a broad spectrum of political and cultural thought. Indeed, there were different types of cultural conservatives – those like Waugh and T. S. Eliot who were pledged to defend the immutable values of an elite minority against the indiscriminations of popular culture, and those like George Orwell, Richard Hoggart and the so-called 'Angry Young Men' who were interested in preserving what they perceived to be real or traditional working-class values against the blandness of commercial culture.

While their political and cultural perspectives may have differed, the threat of mass culture provided common cause for concern. George Orwell's '*Coming up for Air*' (1939) – although written before the outbreak of war – shows this clearly. On the scourge of the emerging American-style milk bars he writes:

'There's a kind of atmosphere about these places that gets me down. Everything slick and shiny and streamlined: mirrors, enamel

An advertisement for Bakelite Ltd. introducing the notion of 'good design' into the promotion of plastic products in Britain. *Design UK* April 1952.

and chromium plate whichever direction you look in. Everything spent on the decorations and nothing on the food. No real food at all. Just lists of stuff with American names. Sort of phantom stuff that you can't taste and can hardly believe in the existence of.'[5]

Although not mentioned by name, 'everything slick, shiny and streamlined' could cater as a description of plastics, while images of modernity and artificiality certainly imply the use of plastics. In any case the overall connotations are clear: that such products (streamlined, shiny, modern) are signs of American decadence and a lessening in cultural and aesthetic standards.

This deep suspicion and hatred of mass culture did not directly affect the bulk of the plastics-buying population – the so-called rank-and-file – who for the most part welcomed their new consumer freedom. It was to have a serious effect, however, on the judgment of key Establishment bodies, the official Arts bodies: the BBC, the Arts Council and the Council of Industrial Design. As well as consolidating the institutional and elitist nature of cultural attitudes in Britain, these bodies acted as guardians of the nation's cultural heritage and as such were seen to be the official educators and arbitors of cultural or artistic value and 'good taste'. In this capacity the Arts Establishment was also charged with the defence of elitist 'high' culture against the vulgar inroads of commercialised mass culture.

It was the Council of Industrial Design's task to educate the nation in matters of design. It perceived its role as one of paternalist adviser, influencing and guiding public taste and persuading manufacturers and consumer alike of the financial, moral and social advantages of 'good design'. Its stance was, significantly, one borrowed from past decades, a mixture of the Arts and Crafts emphasis on quality and craftsmanship, with the Modernist tenet 'form follows function'. There was also a great sense of morality and 'good manners' injected into the Council's ideas. Elements such as 'purity of purpose' were emphasised, as was the sense of moral earnestness and the belief in design as a force for social betterment. Like other cultural conservatives the Council felt that many post-war developments represented a fundamental threat to all that was quintessentially British about Britain, championing the English virtues of tradition and craftsmanship as bulwarks against American manufacturing and selling techniques.

In 1952, Alec Davies the editor of *Design*, the Council's own magazine and most effective public voice, wrote:

'American mass-production methods are hardly appropriate to the makers of say, Staffordshire bone china, Yorkshire woollen cloth, Walsall leather goods, or London-tailored men's wear. They and many other British industries depend for their existence on our tradition for quality, which never existed in the same form in America because there has never been an aristocratic background to set such standards – but always an insatiable demand for goods, so that speed of production became the first priority in American industry.'[6]

The hostility directed towards the streamlined features of American styling, and the American selling techniques that accompanied them, were related to an instinctive distrust of publicity and marketing and a lingering suspicion of commerce that had characterised British culture for the last half century.

Design in plastics obviously came under much scrutiny from the Council, it being a modern mass-produced material and, as such, vulnerable to the evils and excesses of American styling. As Paul Reilly, the Chief Information Officer of the CoID, explained to *Design* readers:

'The temptation to adorn a moulding, to imitate carving or to impress a stylistic cliché, such as the 3 parallel lines is certainly hard to resist when one considers the extreme ease of reproduction in plastics. It is in fact this very ease of processing which so often leads to trouble and to that reputation for tastelessness which clings to certain categories of plastics products; this ease of shaping, of giving a texture (too often an imitation of some natural grain) and of colouring demands great discrimination on the part of a plastics manufacturer or designer. The temptation to exploit the spectrum is considerable and much damage has been done to plastics through gaudy, or, worse still, sentimental colouring.'[7]

By and large, the practice of decorating plastics to imitate wood, ivory and other natural materials had died out with the progress made by the plastics industry during the war and with the introduction of revolutionary new plastics with important and unique qualities of their own. Small amounts of this type of imitative decoration continued to be produced, however, and was indicative of a continuing resistance in Britain to modern materials and a continuing reliance on the comforts of tradition and familiarity. Most plastics design, having survived the disasters of the immediate post-war years, was modelled on imported objects employing the glossy brashness in colour and line that confidently echoed the 'juke-box' culture of American youth which naturally was frowned upon by the British design hierarchy.

Two plastic clocks from the pre-war period representing the kind of products that the 'good design' campaigners considered to be of inferior quality.

Criteria called upon by reforming design critics included the qualities of 'honesty' and 'purity'. As a machine-made material plastics was called upon to reveal itself as such, no decorative techniques could be employed that suggested the article was anything but cheap, man-made, mass-produced and, above all, plastic. In the words of Paul Reilly:

'The designer of plastic mouldings should never lose sight of the fact that his material becomes viscous and 'flows' at some stage in its processing . . . A plastics article must therefore depend mainly on its colour and form. It should make up for the synthetic or dead quality of its substance with vitality of line, which is never impossible with a material that can be made to flow. The designer in plastics should strive to capture this previous plastic state of his material in the form of his finished product . . . too many plastics articles on sale today are

Melamine cups and saucers from the 'Fiesta' range designed by Ronald E. Brookes for Brookes and Adams. England 1961.

A 'castellated' plastic condiment set from the post-war period. A classic example of what was considered to be 'bad taste' in plastic design.

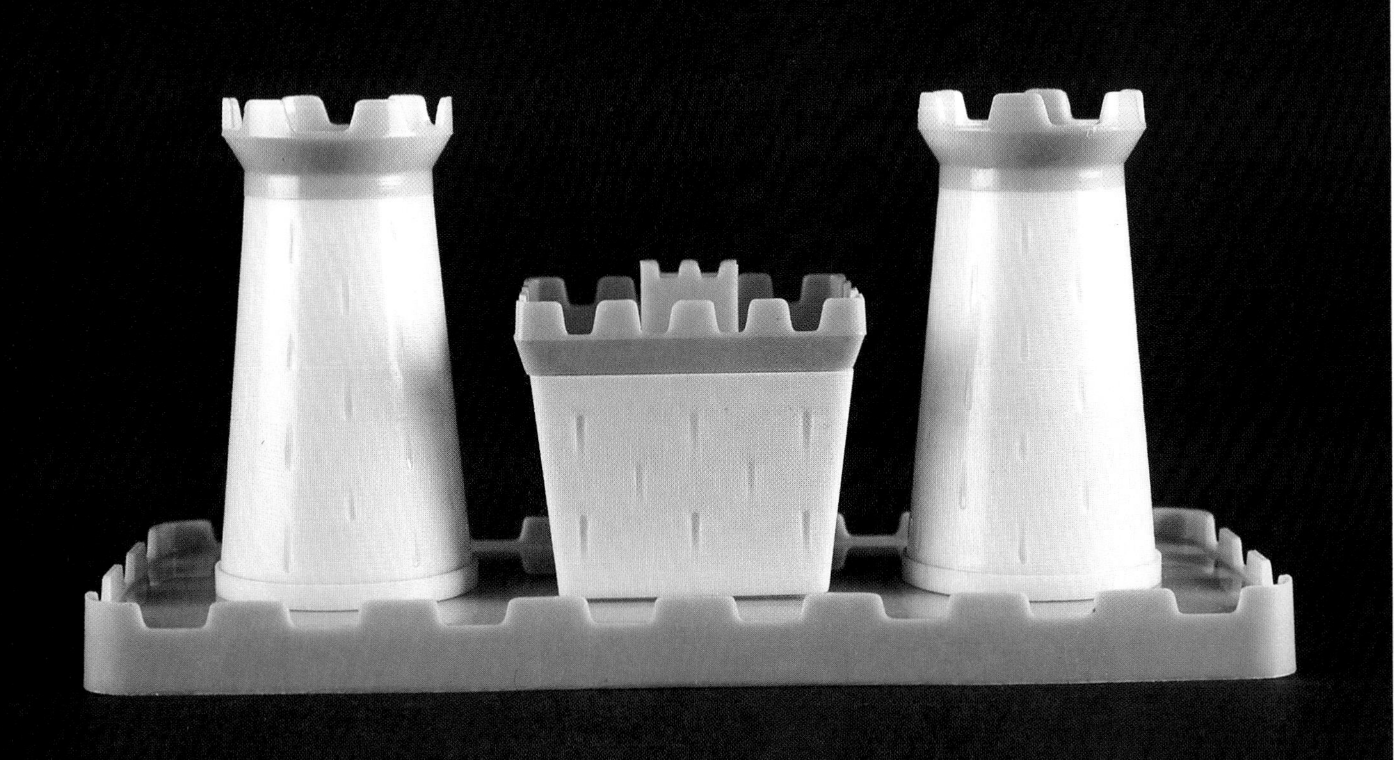

constipated little objects which deny their former free movement, and what they lack in line is wrongly assumed to be compensated for in detail. Detail in a moulded product should be used with reserve for there must be no sale under false pretences. It should never be suggested that midnight oil has been burned in hand carving, or labour spent in individual engraving of a product which is in fact mouldable by the million.'[8]

The emphasis placed on traditional craftsmanship and quality was stressed by the Council. The plastics industry was only too aware of its lack of design heritage; a relatively new material with no past tradition, plastics did not fit well into the complex web of historicism and traditionalism surrounding other well-established industries in Britain. Forced into a precarious position by the 'good design' pronouncements of the Council, the plastics industry attempted to justify its product by association with age-old tradition and craftsmanship.

The most apparent examples of this appear in the advertisements put out by the industry in which they tried to entice their customers by conjuring up images of traditional arts and quality craftsmanship through references to such old masters as diverse as Chippendale, Beethoven, Rembrandt and Shakespeare, and to craftsmen of ancient Rome and China.

The concept of 'good design' was also taken up by the plastics industry, however, and was evident in much of the publicity material it put out. Although much of this emphasis could be seen as a result of the CoID's own conservative attitude, and was often stamped with the same bogus tone of self-righteous morality, it must be stated that

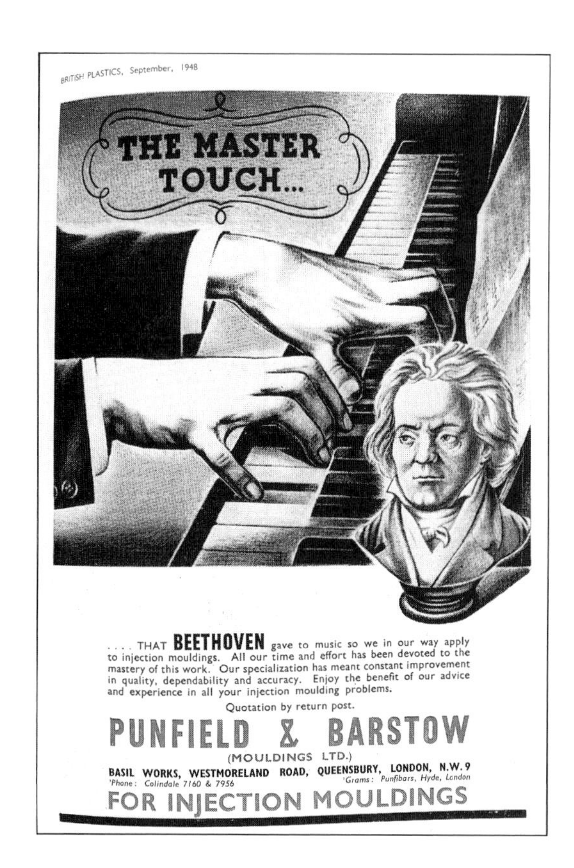

An advertisement for Punfield and Barstow injection mouldings. *British Plastics*, September 1948.

in the case of the plastics product itself it was an emphasis that worked to some advantage.

This emphasis did not encourage adventurous design in plastics, however, the articles most praised by design critics tending to be rather simple cups and saucers, salt and pepper cellars or buckets. However, it did allow for a high critical standard in design to be set. While this stifled much of the more adventurous and colourful qualities of plastics it was able to counteract the production of goods that were either technically ill-conceived or chemically unstable, which was characteristic of so much of the immediate post-war output.

One of the more successful outcomes of this crusade was in the setting-up of design studios within the various firms of the plastics industry, one of the earliest and most outstanding examples being that of British Industrial Plastics Ltd. G. H. Glassey, the managing director, was quick to see that the manufacturer, the trade moulder, as well as the end-user needed educating and advising on the proper use of plastics before their true potential could be realised. In 1947 a Design Advisory Service was set up with two divisions – one dealing with mould design, the other with product design – and was staffed by trained designers whose experience was available to anyone who wanted a product made in plastics, whether a BIP customer or not. The Design Advisory Service team were under instructions to specify the most suitable material for the job and not necessarily those produced by their company.[9] In 1951, a Product Design Unit was set up under A. H. Woodfull to continue and extend this service.

Many of the products designed by the Product Design Unit were used as illustrations in promotional material where, perhaps under influence from the Council, emphasis was placed on the design process. On examination many of these products display little adventure in design and seem consciously to be avoiding the more confident, American-inspired influences in shape and colour. They were, nonetheless significant improvements technically, if not aesthetically, on products produced by the British plastics industry in previous years.

The Council's reinforcement of the myths of traditional and rural values did little, however, to shift the image of plastics in the general public's view. By continuously emphasising the merits of the Arts and Crafts tradition for quality and workmanship, design critics were undermining any favourable reception to plastic products, while the overall reactionary views of the Arts Establishment served only to exacerbate the hostility towards plastics that had existed before the war. Although the Council's emphasis on 'good design' and the efforts of some of the larger plastics companies to react responsibly against the deluge of ill-conceived products did improve the quality of output to a large extent, there continued to exist an apparent lack of confidence and reticence in plastics design in Britain in this period.

In conclusion, plastic products in Britain suffered, in these years, from the entrenched traditionalism of the Estabishment bodies which, fearful of being trodden underfoot by the rush of mass-cultural change, favoured a tentative and highly judgmental approach towards the new goods. As a result – with only a handful of exceptions – British plastics from the 1950s failed to exhibit the same verve and grasp of modernity that can be found in parallel products emerging from other European countries such as Italy, Germany and Sweden. The reasons for this, as we have seen, are more cultural than technological.

Claire Catterall, 1989.

Notes

1. See A. Allcott. *Plastics Today*. London 1960, p.28
2. In the stampede to cater for the consumer market, cut-throat competition led to massive cost-cutting which meant that little regard was paid to the suitability of plastics for specific applications. See R. D. Russell. 'What is wrong with plastics design?' in *Art and Industry*, Oct. 1948, pp.148-49
3. Dick Hebdige. 'Towards a Cartography of Taste 1935-62' in *Block*, 4, 1981
4. Evelyn Waugh, *The Ordeal of Gilbert Pinfold*, London 1957. Also quoted ibid.
5. George Orwell, *Coming Up for Air*, London, 1939. Also quoted ibid.
6. Editorial, *Design*, 42, (Aug. 1952)
7. P. Reilly. 'Pitfalls and possibilities of plastics design' in *Design*, 17, (May 1950)
8. Ibid.
9. John Vale. 'Designing for Moulded plastics in the post-war period' in *From Spitfire to Microchip Studies in the History of Design since 1945*, London 1985

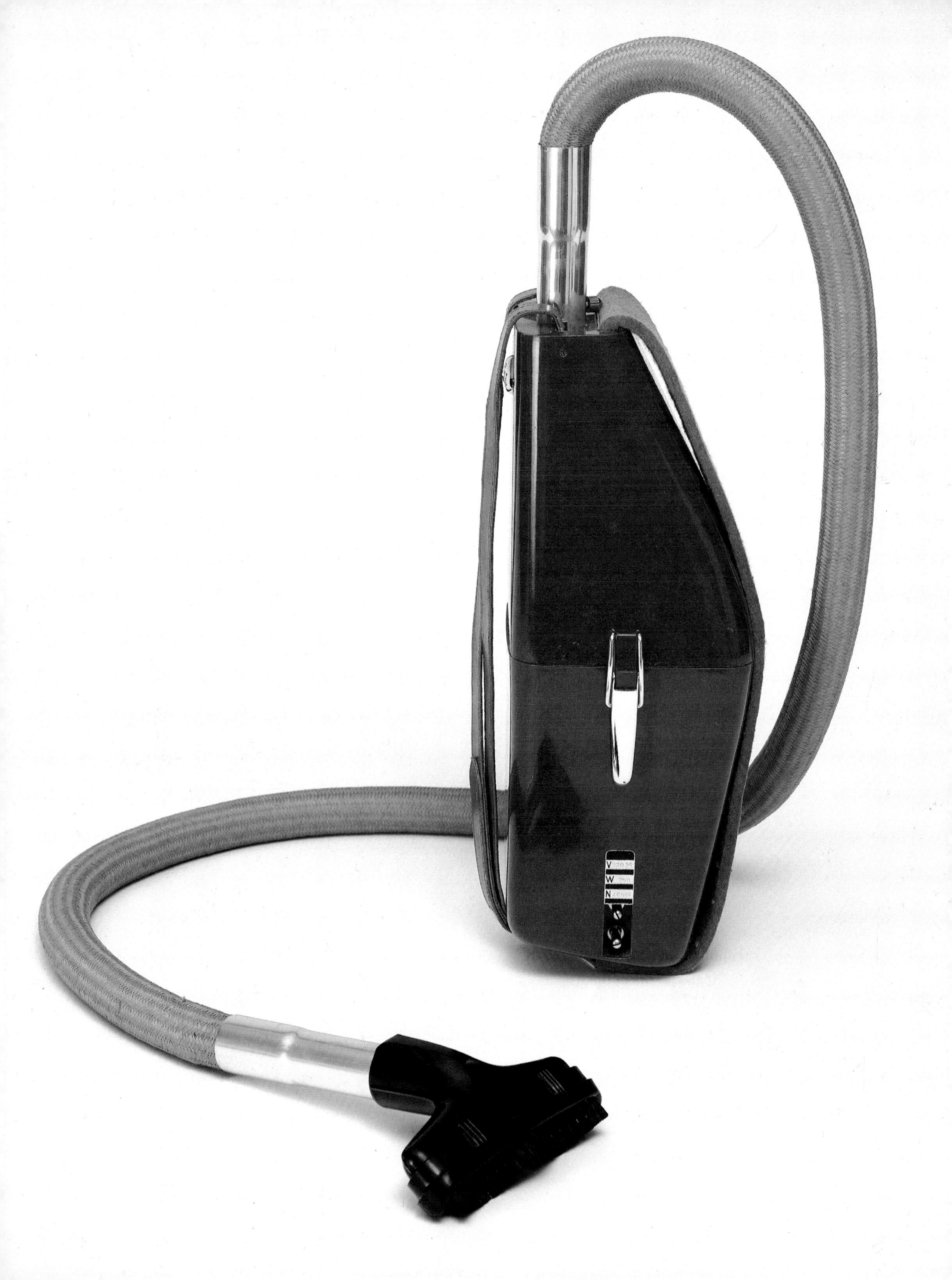
V
W
N

THE ITALIAN WAY TO PLASTICS

Giampiero Bosoni

The 'Spalter' vacuum cleaner, made of nylon, designed by Achille and Pier Giacomo Castiglioni for REM. Italy 1956.

The choice to analyse the development of Italian plastics from the end of the Second World War to the 1960s and subsequently to relate this phenomenon to the unquestionable success in those years of what came to be defined as 'Good Italian Design', confirms the need to contextualise the historical reading of the design of an industrial product with reference to its most significant components.

This is the reason why a material, in this case plastics, constitutes an essential point of reference in an understanding of the historical and geographical coincidence of events in the scientific, economic and social fields which were quite unusual in the Italian panorama, and otherwise difficult to comprehend.

The Italian Plastics Industry before 1946

Nearly fifteen years after the industrial production of the first phenol resins (bakelite, 1910), and some fifty years after the creation of the first commercial nitrocellulose resins, the Italian industrial sector launched the first entirely national production in this field. The first plastic to be produced in Italy was celluloid. It went into production in 1924 in Castiglione Olona (Varese) under the auspices of the Società Italiana Celluloide (an industry which we will meet again many years later under its current name of SIC-Mazzucchelli). After this initial experience and within the space of a short period of time, a number of firms grew up rapidly in this sector and subsequently merged to form the Galalite consortium.

At the beginning of the 1930s, the principal plants producing plastics included the Società Italiana Resine (in Sesto San Giovanni), the Società Monti & Martini (Melegnano) and the Società Nazionale Chimica (Castellanza) which was subsequently taken over by Montecatini. In 1921 this large national industry, which under the auspices of Giacomo Fauser[1] (whose process of fixation of atmospheric nitrogen it was to adopt) initiated a close collaboration with the Universities, understood, if somewhat belatedly, the potential importance of these new materials.

In fact, Montecatini, already responsible for the supply of raw materials indispensable for the production of the new synthetic resins (phenolic and ureic),[2] began to exert its influence in this sector around the middle of the 1930s. A clear sign of its influence was the substantial national increase (75%) in this sector in Italy, which however lagged a long way behind the United States (230%) which had started with quite different levels of production.

Between 1935 and 1936, imports became subject to a system of governmental authorisation which created notable difficulties for the new industries. Adopted in part in response to protectionist measures and to sanctions against Italy adopted by a number of countries, this system edged the Italian economy towards a policy of complete autarchy.[3]

But if on the one hand, as many observers have noted, this autarchy caused the disappearance of the Italian industry from the lively international panorama of continual scientific and technical innovation, on the other hand it is

impossible to ignore the fact that in Italy science and technology assumed an increasing importance in connection with the development of its imperial, autarchic and war requirements. The first congress for the development of industrial autonomy[4] was a highly significant pointer to this situation. Apart from its autarchic theme it highlighted for the first time the problem of industrial research. And it was precisely in this period that Giulio Natta (a figure who was central to the plastic sector in the post-war period) commenced his research activities in industrial chemistry, inevitably under the direct influence of the policy of autarchy.

In 1924, when he was only twenty-one years old, Natta graduated in industrial chemistry from the Milan Polytechnic. At the age of twenty-four he was already a lecturer in general chemistry. In 1928 he patented the first internationally important processes for the production of synthetic methyl alcohol. From 1933 to 1937 he taught at a number of Italian Universities (Pavia, Rome, Turin) and then, in 1938, he returned to the University in Milan as Professor and Director of the Industrial Chemical Institute of the Polytechnic, a post which he was to maintain until his death in 1979.

The most important research carried out in the spirit of political autarchy took place in 1938 when IRI and Pirelli decided to launch the production of synthetic rubber in Italy. Professor Natta who was appointed a consultant of the Società Anonima Italiana Gomma Sintetica. Subsequently, while working on the reproduction of butadene, Natta was the first to achieve the separation of butene 1. As a result of his research this process of extractive distillation was incorporated for the first time anywhere in the world in a plant based in Ferrara. Natta himself pointed out the stimulus which autarchy had on his research activities in one of the numerous interviews which he gave when he was awarded the Nobel prize for chemistry in 1963:

'Autarchy has stimulated many chemical scientists to search for new materials, but many abanadoned that research in the absence of political motivation. I wasn't interested in politics but in research and

Left: Stacking basins, baby baths and buckets designed by Roberto Menghi for Moneta Smalterie Meridionli. Italy 1955.

Below: A polyethylene bucket designed by Roberto Menghi for Moneta Smalterie Meridionli. Italy 1955.

thus when the political need for autarchy came to an end I continued to search.[6]

In this same interview Natta pointed out another important factor in the autarchy which had acted as a stimulus for his influential research; the close collaboration which he had had with the industrial sector and the resulting possibility of examining strictly practical and functional problems.

However, strictly from the point of view of production, the war effort in Italy – largely as a result of a series of economic and governmental failures – constituted a major obstacle to the realisation of the ambitious development programmes in the field of plastic resins. This is in contrast to the situation in Germany, Great Britain and, particularly, in the USA, where the production of synthetic materials experienced an extraordinary growth (from

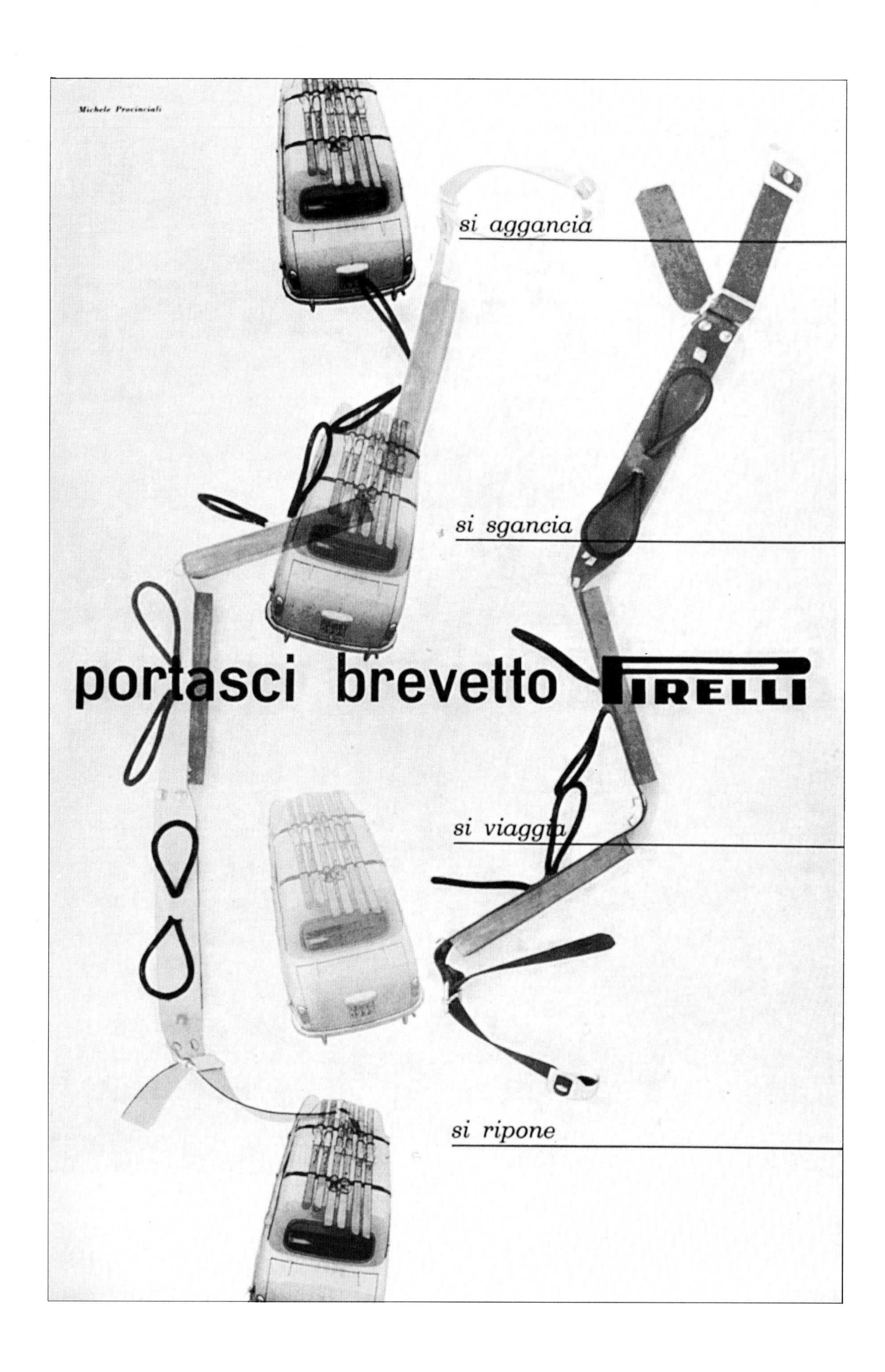
Michele Provinciali
si aggancia
si sgancia
portasci brevetto PIRELLI
si viaggia
si ripone

1939 to 1944 production in Germany rose from 100,000 to 250,000 tons; Britain virtually doubled her production; and in the USA it rose from 106,000 to 410,000 tons).

Undoubtedly the exceptional development of the industry in the USA was due to the spread of a better and more precise awareness of the practical potential offered by plastic materials, particularly where war production was concerned. After Pearl Harbor the American industry was called on to respond to the requirements of a major production programme and it may be of some interest in this context to examine the motives for its decision to employ plastic materials on a wider scale: in the first place the time limits within which the armed forces' requirements had to be met were too often incompatible with the necessary processes involving wood and metals; on the other hand it was necessary to accelerate production and at the same time to reduce the work-force. In the end many rare and precious materials had to be replaced. Already after the first year of the war it became clear that the use of plastic materials made it possible not only to substitute wood and metals in many of their applications and to achieve substantial economies with regard both to time and to materials, but also how many of these offered precise technical advantages over the materials which had been previously employed, suggesting quite novel applications.[7]

This analysis of the development of the war industry in the U.S.A. is vital to the understanding of the origins of a large part of the innovations of a technical, but in particular, an applied nature which were expressed after the war in world industry and especially in Italy. In the first place one shouldn't forget the importance which these new materials had on the formal researches undertaken during this period by the best-known American designers (Eames, Nelson, Loewy) who worked not only as part of the war effort but also, after the war, were to influence Italian design with the models which had been yielded by their research. In the second place, we should remember that the launch of the post-war redeployment of Italian industry was largely the result of the high level to which the American industry had developed. For this reason it is worth noting that during the war years, new processing methods were studied in the USA including, for example transfer moulding, low pressure moulding for the stratified plastics and injection moulding; new products such as melamine and allyl resins, nyinilidene chloride, silicones and polythene were also developed, without the excellent electrical and chemical resistant properties of which the invention of radar, the unproved preservation of weapons and equipment, and a series of other important applications too lengthy to list would not have been possible.[8]

By the end of the war the American industries were acutely aware of the advantages and implicit possibilities of plastics; precisely for this reason the redeployment of industry from military to civilian application was a mark of progress rather than of a standstill.

An advertisement for 'portaski', a Pirelli patent, designed by Roberto Menghi, with the engineer Barassi, for Azienda Pirelli Monza. Italy 1950.

The Post-War Italian Plastics Industry

The petro-chemical industry was also poised to start its boom years in Italy, when the country was wearily picking itself up from the losses and ravages of war. The new-born plastics industry was to be one of the most extraordinary industrial developments on a global basis of this period. A number of unusual events in the Italian context were to have a special impact on this sector. On the one hand, the scientific research undertaken by Giulio Natta and his team in the Polytechnic in Milan achieved a position of pre-eminence through the discovery of polypropylene. This was to be acknowledged by the recognition of this illustrious scientist by the award in 1963 of the Nobel prize.[9] This research, which had come to fruition in its crucial stages in collaboration with Montecatini, the future Montedison, allowed the great Italian chemical industry to obtain the international copyright of this new material under the trade-name of Moplen,[10] a name which was to become synonymous in this period with plastics. In the economic-industrial field, Moplen resulted in the emergence of Montedison into the plastics sector at a world level, with the subsequent sale of plant and copyrights abroad, thus overturning its backward position to reach the peak in this sector. Last, but not least, through international renown, the phenomenon of 'Good Italian Design' played an important role in the consolidation of the Italian image abroad. It was thanks to developments in the plastics sector that Italian design found a confirmation of those very social values which many attempted to make the basis of the industrial product.

In the period immediately after the war the task facing the chemical industry was of primary importance and the greatest efforts were initially dedicated to the reconstruction of the basic production plants which had to be rapidly expanded to meet the growing demands.[11] Fortunately the factories producing plastics and synthetic resins industry survived destruction and dismantling. Manufacturing plants had been damaged to some extent but in general the production potential remained intact thus establishing the basis for a revival which was more necessary than ever, both because of the global orientation towards new materials and because of the scarcity of the more costly traditional materials for which the former were to act as easy and economical substitutes. The growth of our industry was rapid if somewhat disorderly; the substantial American aid was for the most part poorly utilised and in a typically Italian way opposed even the most cautious proposal to follow a programme. However, in this situation another characteristic typical of the Italian industry was to take over and redress the balance, namely the lively entrepreneurial spirit of many small businesses which early on became the object of attention from abroad. The magazine *British Plastics*[12] explained:

'It's obvious that the Italian industry will rapidly make further progress. The Italian technical triumphs of the pre-war period have not been forgotten despite the occupation and the subsequent break-down of the industry. Technicians have a natural ingenuity which allows them to improvise with sub-standard materials and equipment; perhaps our complete familiarity with the best quality has left us short on ingenuity. My impression is that the Italian producers of plastics are not a long way off being able to compete seriously with English products in the world markets.'

This forecast was to prove accurate, for within the space of a few years Italy reached fifth place on the world production scale.

A pile of plastic containers designed by Roberto Menghi for Azienda Pirelli Monza. Italy 1959.

From 1946, the year in which the manufacturing[13] companies which formed part of the Associazione Italiano delle Materie Plastiche already numbered 237, there was an almost geometrical development from one year to the next. By the early 1960s this was to bring the Italian industry up to second place in the rate of growth on a world level.[14] In actual fact from 1950 to 1960 a substantial increase in internal consumption (825%) and a quite remarkable leap in exports (4850%) were recorded, while imports, overtaken by exports from 1954, increased by a notable, but not exceptional, 900%. It is important to remember in this context the influence which the engineering industries exerted, specialising in the construction and perfection of machine tools and above all in responding to the needs of smaller markets like ours, demonstrating a far-reaching vision confirmed by the excellent capacity for exporting practised in all, including the most advanced world markets.

On the crest of this extraordinary development of the sector, other chemical and industrial business enterprises followed the example of Montecatini endowing plastic materials with the qualities to inspire wide interest and extensive consumption.

The Question of Aesthetics and the Plastic Product

Among these industries a number were to distinguish themselves through the maturity of the aesthetic value of the applications of plastics in their products. However these isolated cases were insufficient to prevent the overall application of plastics as practical and economic substitutes for natural materials. Thus, in considering the

Entrance to the first International Exhibition of the 'Aesthetic of Plastics' at the 34th Milan Trade Fair, designed by Alberto Rosselli. Italy 1956.

design of industrial products in plastic, one should take account of the enormous quantity of badly designed products which were useless and vulgar. These should be remembered and studied if only because of the extent of their production and the influence on the aesthetic and the social culture of an entire epoch which resulted from the bizarre forms and unusual colours which it was possible to achieve with this new material. Thus one should speak of domestic objects fashioned from plastics which took the place on the market stalls, of enamelled iron and pottery and thus the counter-productive but inevitable 'thickness battle' carried out by small and medium-sized laboratories from which was generated the discredited image of synthetic resins as worthless materials.

However there was no shortage of attempts to qualify plastic materials for their intrinsic applicational qualities; from the early 1950s several companies took advantage of the continual collaboration with recognised designers, in a number of cases even to the point of deciding to place them in charge of the projects office.

A personality like Gio Ponti, an indefatigable promoter of continuing contacts between industry and the architectural culture, formulated the proposal – in 1955, at a congress organised by Montedison – of an international exhibition on the aesthetics of plastic materials. Sponsored by the magazines *Stile Industria* and *Materie Plastiche* the exhibition was held the following year in the Fiera Campionaria in Milan, presenting a selection of some 160 objects of international origin. The magazine *Stile Industria*,[15] without doubt the best-qualified publication on industrial design dedicated an entire number to this subject, attempting to suggest through the use of articles such as 'Notes and indications for moulding' a number of useful pieces of information for those working in the field.

In addition, the Compasso d'Oro award was constituted as an important means of giving credibility to products using plastic

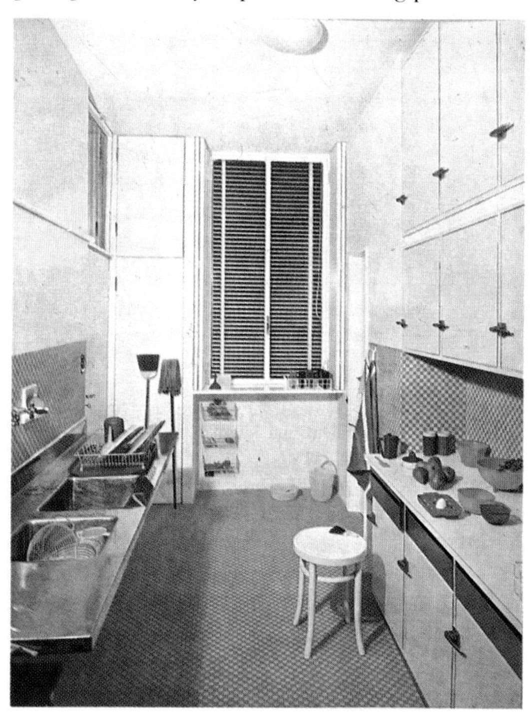

Plastic objects in a kitchen illustrated in *Domus*. Italy 1954.

materials. In fact it is possible to find fascinating articles in specialised magazines such as 'Materie Plastiche' or 'L'Industria Italiana dei Poliplastici' describing this prize in a way not seen in magazines directed to the world of design. These articles were particularly interesting insofar as the authors dwelled on the technical and functional aspects of the application of the different plastic materials used for those products which won prizes or were selected.

Virtually all the designers of this period had the opportunity to manufacture products in which plastic materials appeared, at least to some extent. But the best-known cases of exemplary relationships between design and industry were without doubt the collaboration of Roberto Menghi with Pirelli and Moneta Smalteria Meridionali, the partnership between Marco Zanuso and Arflex as far as the use of foam rubber was concerned; and, above all, the case of Gino Colombini and Kartell which represents the most successful example of the fusion of technical and formal research.

The cover of *Stile Industria* no. 7 1959 which was dedicated to the International Exhibition of the 'Aesthetics of Plastics'.

In Italy plastics found their first widespread use in domestic products in the early 1950s. In this period Kartell was founded by Giulio Castelli (in 1949) and it almost immediately became the best-known reference point for the birth of Italian contemporary industrial design of that period.

Castelli had specialised in chemical engineering under the guidance of Giulio Natta, but his real interest and experience in this field came to him less through his university specialisation than through his father who, even before the war, had set up a laboratory producing Bakelite objects.

Kartell first started its activity in the sector of car accessories, and achieving a certain notoriety with the production of a ski-holding rack, a Pirelli copyright, designed by Alberto Menghi with the engineer Barassi – a technician with Pirelli Monza and a little later to be one of the founders of Arflex.

In the early 1950s, Castelli imported the first polythene bowls and containers from England which, as a result of their novelty, were priced at prohibitively high levels. For this reason Kartell started producing their own plastic household articles in 1953 which also marked the beginning of the collaboration with Gino Colombini, initially freelance and subsequently as one of the project-design team. Colombini, who had trained in Albini's studio, was to be the designer of almost the whole range of Kartell's products from 1953 to 1962. In this period Colombini created more than 200 new household articles and was awarded the

'Compasso d'Oro' on four occasions and received 54 merit nominations. This unusual, and at that time unique, relationship of a designer working as a staff member within a company, allowed Colombini to develop a working-relationship both with Mr Viviani, an engineer in the moulding department and with Mr Neri, the engineer in the production section, which proved highly stimulating for the creation and realisation of a variety of products. However, it was undoubtedly with Castelli that this relationship was most powerfully bonded not only because Castelli had the greatest influence in the management of the firm but also, and above all, because as a technician he was able to steer an even course between the proper formal experiments which Colombini proposed and the schematic, and sometimes conventional, methods favoured by the technicians. In 1955 Kartell became a producer, for the Società Mazzucchelli Celluloide S.p.A. of Castiglione Olona (SAMCO), of household products manufactured by injection techniques.

From 1956 to 1961 the information catalogue 'Qualità' was published by Kartell – one of the first house organs to be published in Italy; after the third issue it was transformed into a fully-fledged magazine with the artistic co-operation of Michele Provinciali. The development of Kartell saw the birth in 1958 of two new important divisions: the 'laboratory for technical articles' division and one for 'lighting fixtures' which involved collaboration with new designers like Asti and Fabre, Castiglioni, Menghi, Monti and Zanuso.

However it was to be the polyethylene chair 'K4999' (the first chair manufactured entirely in plastic, designed in 1964 by Marco Zanuso which marked the peak of the company's activity and indeed was to become almost the symbol of the enthusiastic formal experimentation with new plastic materials which this period had aroused. Thus we may conclude that there are fundamentally three major periods, decades more or less, in the evolution of plastics in Italy: the 1950s which included the phase of research and organisation for the industries involved in the application of these materials in various production sectors; the 1960s

Gino Colombini. Colanders, model K1036, manufactured by Kartell SpA. Italy 1959.

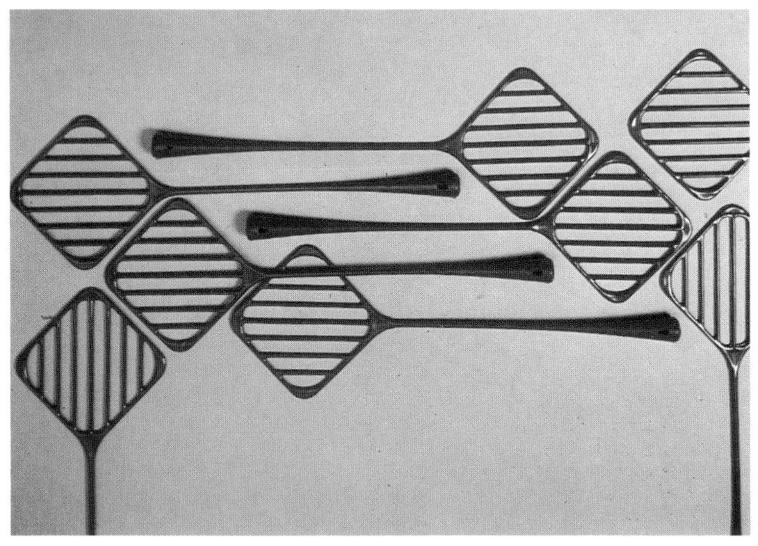

Top: Gino Colombini. Lemon squeezer, model KS 1481, manufactured by Kartell SpA. Italy 1958.

Gino Colombini. Carpet-beater, model KS 1465, manufactured by Kartell SpA. Italy 1956.

which saw an enthusiastic, and often uncontrolled, use of plastics which coincided, on the one hand, with this material's novelty and on the other, with its low cost; and finally the 1970s which experienced a crisis in the form of a rejection of 'plastics' and a return to more traditional materials. This situation, together with the energy crisis which substantially increased the cost of both the raw materials and production, today constrains the industry to limit its sales to only those products offering firm guarantees of success. The result of this entire process is a more balanced position of plastics in the production world, and in particular in the household articles market where there is evidence of an improvement in the quality of both the materials and their application.

Notes

1. L. Maino. *Giacomo Fauser. Dodici lustri per la chimica.* Novara 1962
D. Maveri. *Giacomo Fauser.* Novara 1972
2. The production of ureic powders and resins started in 1935 at SIR which worked on a British licence under the trade-mark of BEETLE.
3. Rolland Santi. *Fascismo e grande industria* 1919-1940. Milan. Moizzi Editore, 1977
4. D. Pagani. 'Il primo convegno per lo sviluppo della sperimetazione ai fini dell'autarchia industriale', in *Rivista Marelli.* 1938
5. With reference to his contribution to industrial chemistry before the Second World War, see the information published in *La chimica e l'industria*, June 1943, on the occasion of the awarding of the Royal prize for chemistry to Natta
6. Editorial 'Premio Nobel per la chimica' in *Avanti*, 6 Nov. 1963
7. Andrea De Angelis. 'Le materie plastiche e la difesa militare americana' in *Materie Plastiche.* 1952
8. Ibid.
9. The 1963 Nobel prize for Chemistry was awarded by the Swedish Royal Academy of Sciences to Professor Giulio Natta and the German scientist Carl Ziegler *ex aequo*. Ziegler, the Director of the Max Planker Institute in Mülheim, Ruhr, opened the way to Natta's research with his studies on the polymerisation of ethylene, and in fact the catalyst for the regular linear polymerization of propylene bears Ziegler's and Natta's names
10. In those years, Montecatini obtained a patent for the Moplen resin (isotactic polypropylene), but also the patents for Meraklon synthetic fibre (propylene fibre), Moplefan film (polypropylene film) and Dutral elastomeler. Natta-Montecatini patents for the production of isotactic polypropylene were sold to large-scale foreign petrochemical companies such as ICI (Imperial Chemical industries), Shell Chemicals, the Swedish Esso, Mitsubishi, Sumitono, Mitsui
11. L Morandi. 'L'industria chimica italiana dal 1953 al 1962' excerpt from *La chimica e l'industria*, July 1963
12. E. G. Fisher. 'An impression on the plastics industry in Northern Italy' in *British Plastics*, July 1949
13. While in 1941, manufacturing firms were about 175, in 1947 they had grown to about 500 and in 1967 to 2500
14. Vittorio Diamanti (ed.). *Guida alle materie plastiche.* Milan. Etas Kompass, 1970. pp.292-3.
15. Editorial in 'La prima mostra internazionale dell'Estetica celle materie plastiche' in *Stile Industria*, 7, 1956.

PART THREE

PLASTICS AND POST-MODERNITY 1961-1990

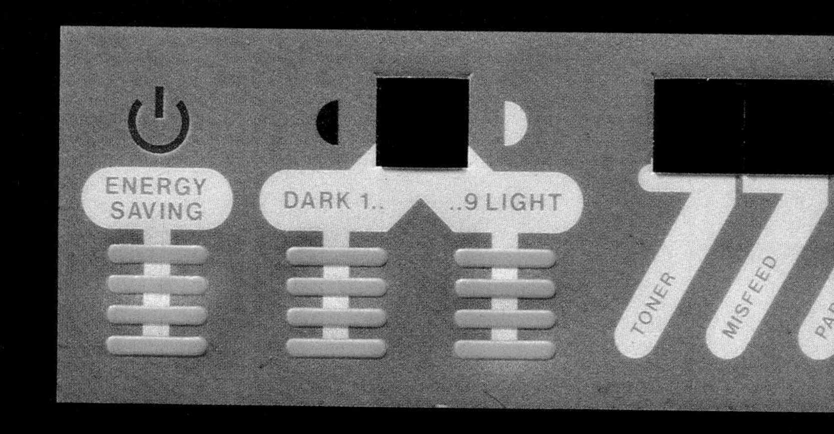

INTRODUCTION

The third, and final, section of this anthology brings together a number of pieces of writing which share the characteristic of being products of what has been called the 'Post-Modern Age'. What this means is, in the most basic sense, that they have all accepted the fact that the monolithic, progressive vision of the Modernists has become obsolete and has been replaced by a much more open-ended cultural system which accepts variation, pluralism and change. While the politics of this much less dogmatic 'movement' are more ambivalent and open to interpretation than those of Modernism – it has been called an 'acceptance culture' which rejects the idea of a moral system – where mass culture is concerned, Post-Modernism has much more to say to us than its high-minded antecedent. Within this new 'zeitgeist' there is, in fact, a blurring of the boundaries between 'high' and 'low' culture and, as a result, plastic products suddenly find themselves in the centre of the picture, capable of incorporating all the meanings that are possible within a post-modern epoch.

Within this new movement the emphasis swings away from production and towards consumption – from, in other words, technology to culture – and the meaning of the object becomes more important than its means of manufacture. In the light of this shift the whole range of possible meanings denoted by plastic products – substitution, modernity, ecological crisis etc. – is acceptable as a mirror of the pluralistic society which engenders them. No longer is there a need to find *the* 'authentic' aesthetic for plastic products but rather a desire to expand the available visual options. While technology – much extended in these years – makes this *possible*, it is culture which makes it *necessary*.

The first two essays in this section – my piece on 'Plastics and Pop Culture' and Reyner Banham's review of the cult film *Barbarella*, entitled 'The Triumph of Software' (from *New Society*, 31 Oct. 1968), both focus on the 1960s in Britain, a decade which acted as a hiatus between the second generation of Modernism and its cultural alternative. Pop represented an outburst of reaction to what it considered to be the stilted voice of Modernism, a spontaneous refusal to accept its, by then, outmoded values. It expressed itself through radical gestures which opposed what was seen as the 'Establishment' culture of the day. As such it anticipated Post-Modernism and many of its proposals were subsequently developed as specifically post-modern themes.

One of the ways in which Pop expressed itself was through the mass-produced material culture of the day and, in this context, it presented the possibility of things incorporating 'expendable' values, whether literally or metaphorically. This was in line with what had come to be called the 'throwaway' culture. In physical terms the new soft plastics – PVC and polyurethane foam – were ideal exponents of this new system of values and they were used extensively as signs of a new relationship between the consumer/user and the physical environment. While my essay looks back at, and examines, the uses of these materials in a number of furniture and fashion projects, Banham's piece focuses on their

David Harman-Powell. 'Nova' tableware for Ekco Plastics Ltd. 1967.

Dress, made of Cardine, designed by Pierre Cardin. Paris 1968.

architectural/environmental implications, stressing the radical new meanings suggested by flexibility and softness.

Both the French writers – Roland Barthes and Jean Baudrillard – represented in this section, approach the subject of plastics as critics of contemporary culture, keen to see how these modern materials function within it. Both are concerned with the 'meanings' rather than the physical properties of plastics. In his essay 'Plastic' (from his anthology *Mythologies*, first published by Editions du Seuil in 1957 and later, in English, by Jonathan Cape in 1972), Barthes wonders at the magical transformation that results in plastic products but feels, ultimately, that this material's inability to divorce itself from the process which forms it renders it 'disgraced' and separated from other more 'genuine' materials. Its real contribution to culture, claims Barthes, lies in its ability to be prosaic rather than luxurious. In his short piece on 'Natural wood, Cultural wood' (from his book *The System of Objects* published by Editions Gallimard in 1968), Baudrillard compares plastics with natural substances, attempting to eliminate a 'hierarchy' of materials and claiming that the new synthetics possess their own characteristics which render them valid in their own terms.

The next three essays in this section – Ezio Manzini's 'Objects and their Skin' (an excerpt from his book *The Material of Invention* published by Arcadia in 1987, and reprinted in *Ottagono* December 1987); Barbara Radice's 'Plastic Laminate' (taken from *Memphis: Research, Experiences, Result, Failures and Successes of New Design* published by Rizzoli in 1984; and Ezio Manzini's 'And of Plastics' (from *Domus*, 666 Nov. 1985) – set out to examine the specific aesthetic qualities that plastics can offer our post-modern culture. Manzini's writings lie at the very heart of this anthology. Suspicious of the attempts, so frequent in Italy in the 1960s, to search for a single aesthetic solution for plastic products, Manzini – a historian of technology – prefers to emphasise the vast aesthetic diversity offered by plastics as a direct result of their ever increasingly complex technological

Printed laminate designed by George Sowden, used on his 'Luxor' cabinet which was part of the first Memphis collection, and manufactured by Abet Laminati. Italy 1981.

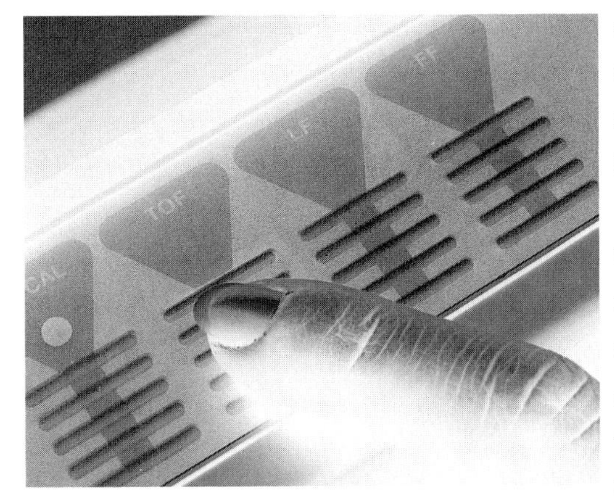

Perry King and Santiago Miranda. Sensitive keyboard for control console of an Olivetti telecopier. Italy 1980s.

possibilities. Above all, it is the shift from form to surface – to the 'skin' of the product – that characterises the new objects, claims Manzini. This new emphasis represents a total break with the ideals of the Modern Movement and the emergence of an alternative way of thinking about both designing and using mass-produced objects: The complexity of plastics manufacture, and the possibility of using compound materials, increasingly 'alienates' us from direct physical knowledge of objects and encourages us, instead, to interact with their surface in a totally spontaneous way. Emphasising the role of plastics as a 'skin' for products allows for the inclusion for such qualities as texture, decoration, luminosity and transparency which help widen the range of possible sensorial encounters with objects.

In their extensive use of plastic laminates in their furniture pieces the designers associated with the Milanese 'Memphis' group put Manzini's ideas into practice allowing the decorated surfaces of their cupboards, tables and chairs to emphasise the role that image, rather than form, increasingly plays within our contemporary culture. Radice's essay describes the mass-cultural origins of this new approach to design. For these designers such an approach meant a liberation from the 'good taste' of mainstream aesthetics.

Manzini's essay about plastics extends these themes. He argues here that the new materials are at the very heart of, and not merely a consequence of, the new cultural context. Their inherent tendency towards what he calls a 'multiplicity of images' and their intrinsic adaptability represent, for him, the pluralism within contemporary culture and in fact make it visible.

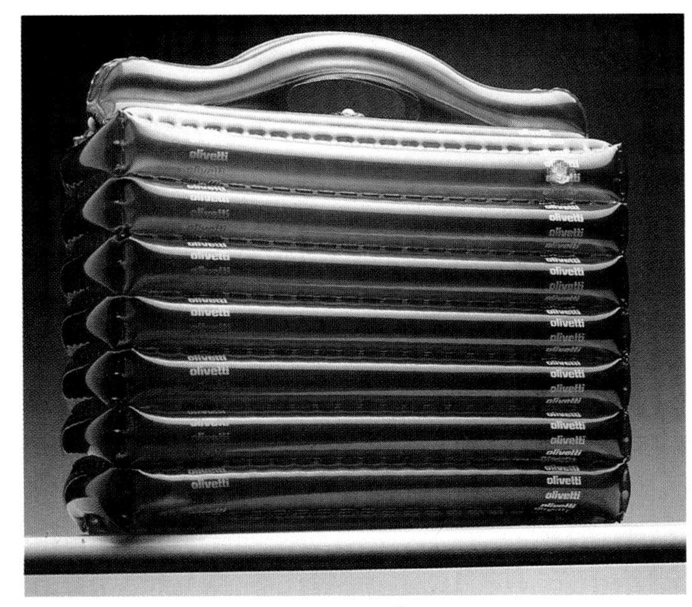

Mario Bellini. Plastic carrying case for his ET 55P typewriter manufactured by Olivetti. Italy 1987.

The final article in this section – Sylvia Katz's 'Plastics in the 1980s' serves to complete the anthology and bring it right up to the present day by showing what is possible today and where plastics are heading in the 1990s. We end, therefore, with a picture of plastics becoming ever increasingly complex and their cultural meanings ever harder to pin down. The challenge that this offers designers is almost without limits.

Penny Sparke

PLASTICS AND POP CULTURE

Penny Sparke

'The aesthetics of Pop depend on a massive initial impact and small sustaining power and are therefore at their poppiest in products whose sole object is to be consumed.' *Reyner Banham*[1]

The material aspects of the cultural movement which emerged in Britain, America and (although to a lesser extent) Europe in the 1960s, referred to by the general name of 'Pop', have, as yet, received scant attention. While much has been written about the striking stylistic metamorphoses which characterised the music, clothes and other 'life-style' accessories of the newly affluent youth markets of that decade, the significance of the recurrence of certain materials in the goods they consumed so voraciously has been scarcely noticed. This essay sets out to fill that vacuum and to show that, albeit within a limited arena, plastics were of fundamental importance to the Pop revolution: They provided a strategy for conveying, through direct physical means, a number of fundamental values embraced by that mass-cultural movement.

The key to the meaning of Pop, as it emerged in Britain in the early 1960s, lay in its direct opposition to the values of the epoch which immediately preceded it. Thus when a new generation of young people set out to demonstrate their beliefs through their consumption choices, its primary intention was to reverse the values of its parents who had experienced the austerity of the war, and the immediate post-war, years. A commitment to permanence was ousted by a desire for constant change, and rigid beliefs were exchanged for more flexible, open-ended ones.

In the objects the new generation consumed this was reflected in a desire for increased expendability and flexibility – whether symbolic or actual. In specifically aesthetic terms this meant a wholesale rejection – expressed consciously by producers and unconsciously by consumers – of the dogmatic ideals of the Modern Movement which still dominated design theory and practice in the 1950s. In essence the theory of Functionalism which underpinned Modernist thinking – and which, in short, required that the outward appearance of an object reflected its means of production – was replaced by a realignment between an object's form and its expressive relationship with the ideals and aspirations of the society into which it was destined to play a part. The desire for universality, therefore, hitherto highly valued by the protagonists of the Modern Movement, was replaced by a need for shorter life-cycles and less monumentality in those objects which made up the material environment of Pop culture. This conformed to, and indeed was largely determined by, the economic exigencies of the mass-

'Sea Urchin' chair, made of upholstered polyether foam, designed by Roger Dean and manufactured by Hille. England 1968.

consumer society and the need for objects to reflect, through their aesthetic and their materials, the culture which both created and sustained them.

Where the design of many of the consumer artefacts of Pop culture was concerned everything focused on two possibilities, namedly an 'aesthetic of expendabiltiy' or an 'expendable aesthetic'. While the former symbolised the 'throwaway culture' the latter was actually disposable. The difference, in material terms, was represented by, on the one hand, a PVC inflatable chair which rejected the notion of fixed solidity and, by implication, the essential formalism of Modern Movement artefacts and, on the other, a paper chair which could be thrown away after a few weeks of use.

The question of disposability was, inevitably, a problem where plastics were concerned but, in symbolic terms, some of the new plastics which were just moving into general use in this decade became, nonetheless, a crucial component of a large number of Pop artefacts. They were the ideal materials to represent several key Pop themes – among them a commitment to technology and the future; formlessness; flexibility; and softness. In addition they offered the possibility of decorated surface which was important to many of the objects of Pop.

Post-war plastics – the results of numerous war-time advances in this area offered many new possibilities to both the designer and the manufacturer of consumer goods. The decade immediately following the war, however, had experienced a down-grading of the plastic product as increased mass-production and, in many cases, shoddy manufacturing had caused plastics to be

Top: Joe Colombo. Chair, model 4867, originally moulded in ABS by Kartell SpA. Italy. 1968.

Vico Magistretti. 'Selene' dining chair, made of ABS and manufactured by Artemide. Italy 1966.

thought as 'inferior' materials. The explosion of plastic products into the mass environment in the years 1945 to 1960, and the growing reaction to them, had threatened to undermine the full significance of the technological advances that had been made in this area. In response to this attitude a handful of designers and manufacturers in Italy, Great Britain, Scandinavia and elsewhere, set out to restore the reputation of plastics to the position they had occupied before the war ie; as eminently acceptable alternatives to more expensive materials and as *the* appropriate materials for the new products created by advances in technology; to imbue plastic products with a concept of 'quality'; and to establish a 'high cultural' role for them. In order to bring about this recuperation the principles of the Modern Movement were brought back into focus and, in the late 1950s and early 1960s, the Italian designers Marco Zanuso, Vico Magistretti and Joe Colombo produced chair designs for the Kartell and Artemide companies which were highly formal studies demonstrating the way in which plastics could be used to create high-quality (in both the aesthetic and material sense of the word), monumental objects which could stand alongside the 'classic' furniture pieces of the early century. Their use of GRP (later ABS), and in Zanuso's case of high density polythene, enabled them to produce sleek, shiny, hard chairs which, visually, controlled the space around them. In so doing they helped create what was considered to be an 'authentic' identity for objects made of plastic.

The existence of these sophisticated statements, together with work of a similar status being produced in a number of other European countries, created a new role for

Marco Zanuso. Child's stacking chair in high density polythene, model 4999, manufactured by Kartell SpA. Italy 1967.

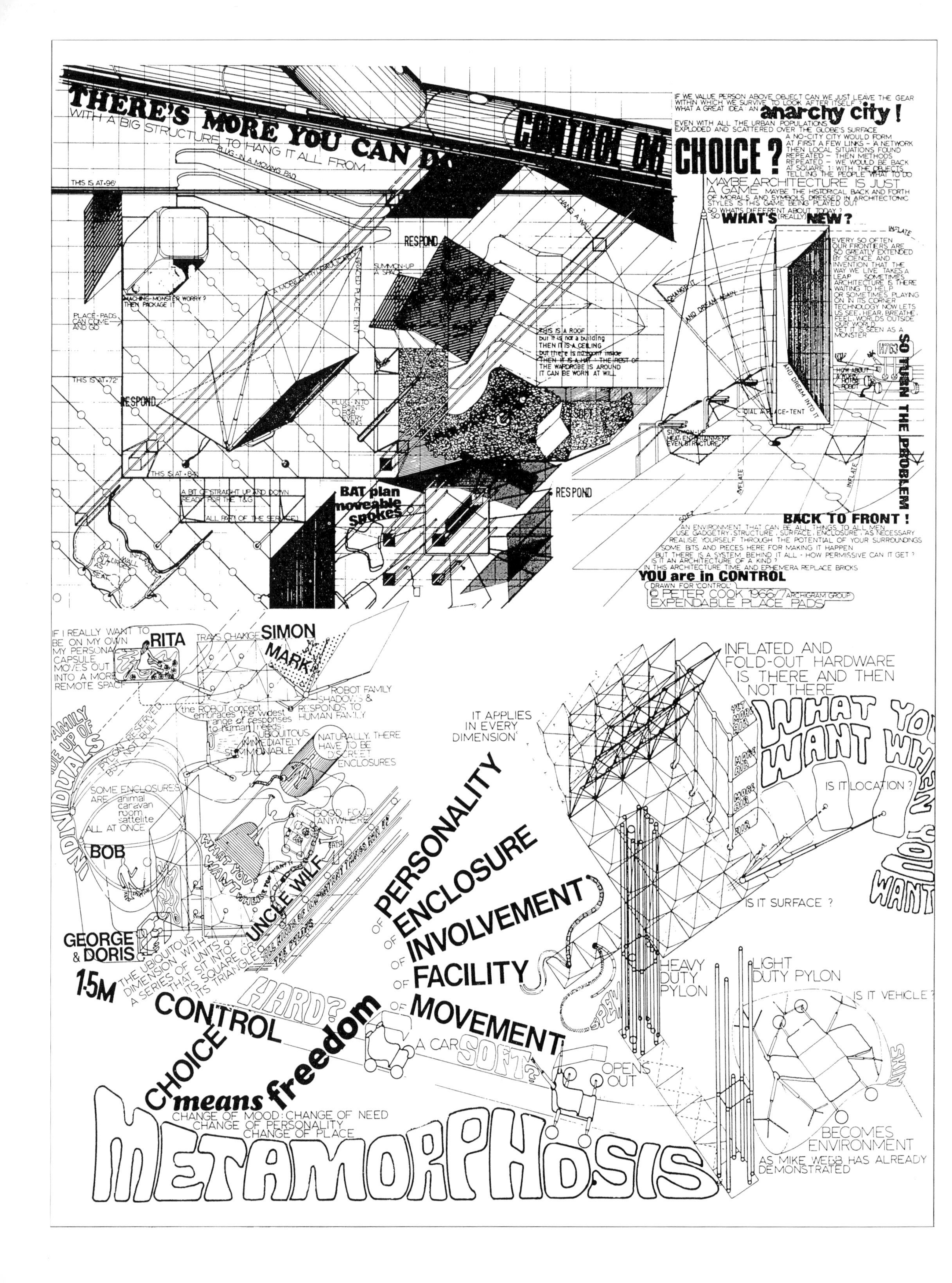

THERE'S MORE YOU CAN DO
WITH A BIG STRUCTURE TO HANG IT ALL FROM
PLUG-IN A MOVING PAD
CONTROL OR CHOICE ?
IF WE VALUE PERSON ABOVE OBJECT CAN WE JUST LEAVE THE GEAR WITHIN WHICH WE SURVIVE TO LOOK AFTER ITSELF ?
WHAT A GREAT IDEA : AN anarchy city !
EVEN WITH ALL THE URBAN POPULATIONS EXPLODED AND SCATTERED OVER THE GLOBE'S SURFACE
A NO-CITY CITY WOULD FORM
AT FIRST A FEW LINKS - A NETWORK
THEN LOCAL SITUATIONS FOUND
REPEATED - THEN METHODS
REPEATED - WE WOULD BE BACK AT SQUARE 1: WITH THE OBJECTS TELLING THE PEOPLE WHAT TO DO
MAYBE ARCHITECTURE IS JUST A GAME. MAYBE THE HISTORICAL BACK AND FORTH OF MORALS AND SYMBOLS DRESSED IN ARCHITECTONIC STYLES IS THIS GAME BEING PLAYED OUT
SO WHATS DIFFERENT ABOUT TODAY ?
SO WHAT'S (REALLY) NEW ?
EVERY SO OFTEN OUR FRONTIERS ARE SO GREATLY EXTENDED BY SCIENCE AND INVENTION THAT THE WAY WE LIVE TAKES A LEAP. SOMETIMES ARCHITECTURE IS THERE WAITING TO HELP OR SOMETIMES PLAYING ON IN ITS CORNER
TECHNOLOGY NOW LETS US SEE, HEAR, BREATHE, FEEL, WORLDS OUTSIDE OUR WORLD YET IT IS SEEN AS A MONSTER
SO TURN THE PROBLEM BACK TO FRONT !
THIS IS AT 96'
THIS IS AT 72'
THIS IS AT 84'
HANG A WALL
RESPOND
SUMMON-UP A SPACE
MACHINE-MONSTER WORRY ? THEN PACKAGE IT
PLACE-PADS CAN COME AND GO
THIS IS A ROOF but it is not a building THEN IT IS A CEILING but there is no room inside THEN IT IS A HAT - THE REST OF THE WARDROBE IS AROUND IT CAN BE WORN AT WILL
PLUG-INTO POINTS FOR EVERY THING
SOFT
CHANGE IT
AND DREAM AGAIN
AND DREAM INTO IT
DIAL A PLACE-TENT
SUMMON-UP HEAT, ENTERTAINMENT, EVEN STRUCTURE
HOW ABOUT A WORK ROBOT
INFLATE
A BIT OF STRAIGHT UP AND DOWN READY FOR THE T&G
ALL PART OF THE SERVICE
BAT plan moveable spokes
SPLURGE OUT AS SEAT
INFLATE WALL
AN ENVIRONMENT THAT CAN BE ALL THINGS TO ALL MEN
USE GADGETRY, STRUCTURE, SURFACE, ENCLOSURE, AS NECESSARY
REALISE YOURSELF THROUGH THE POTENTIAL OF YOUR SURROUNDINGS
SOME BITS AND PIECES HERE FOR MAKING IT HAPPEN
BUT THERE IS A SYSTEM BEHIND IT ALL - HOW PERMISSIVE CAN IT GET ?
IS IT AN ARCHITECTURE OF A KIND ?
IN THIS ARCHITECTURE TIME AND EPHEMERA REPLACE BRICKS
YOU are in CONTROL
DRAWN FOR 'CONTROL'
© PETER COOK 1966/7 ARCHIGRAM GROUP
EXPENDABLE PLACE PADS
IF I REALLY WANT TO BE ON MY OWN MY PERSONAL CAPSULE MOVES OUT INTO A MORE REMOTE SPACE
RITA
SIMON
MARK
ROBOT FAMILY SHADOWS & RESPONDS TO HUMAN FAMILY
the ROBOT concept embraces the widest range of responses to human needs
UBIQUITOUS IMMEDIATELY SUMMONABLE
NATURALLY, THERE HAVE TO BE DISCREET ENCLOSURES
IT APPLIES IN EVERY DIMENSION
INDIVIDUALS
SOME ENCLOSURES ARE animal caravan room sattelite ALL AT ONCE
BOB
GOOD FOOD ANYWHERE
WHAT YOU WANT
UNCLE WILF
GEORGE & DORIS
1·5M
THE UBIQUITOUS DIMENSION WITH A SERIES OF UNITS THAT SIT INTO ITS SQUARE OR ITS TRIANGLE
OF PERSONALITY
OF ENCLOSURE
OF INVOLVEMENT
OF FACILITY
OF MOVEMENT
CONTROL
CHOICE
HARD ?
SOFT ?
means freedom
A CAR
OPENS OUT
INFLATED AND FOLD-OUT HARDWARE IS THERE AND THEN NOT THERE
WHAT YOU WANT WHEN YOU WANT
MORE AIR
IS IT LOCATION ?
IS IT SURFACE ?
HEAVY DUTY PYLON
LIGHT DUTY PYLON
IS IT VEHICLE ?
BECOMES ENVIRONMENT
AS MIKE WEBB HAS ALREADY DEMONSTRATED
CHANGE OF MOOD : CHANGE OF NEED
CHANGE OF PERSONALITY
CHANGE OF PLACE
METAMORPHOSIS

plastic products as conveyors of the principles of 'good design' – a concept which preoccupied countless design critics and practitioners in these years. For the first time plastics earned the reputation of being among the 'quality' materials of the mid-twentieth century.

With the reaction to high culture that the Pop revolution brought with it, however, these objects came to represent one wing of the 'Establishment' culture which, for many, had become increasingly obsolete. Ironically, one of the ways in which opposition to them was manifested was through the creation of more plastic products, this time, however, utilising the new softer materials which communicated quite different meanings from their immediate predecessors.

The introduction, in the mid-1960s, of high-frequency welding as a means of joining together pieces of vinyl chloride which had been polymerised with vinyl acetate (ie; a new version of PVC which was both tough and flexible) made possible the emergence of inflatable structures and objects, thereby offering Pop culture a material which could unite transparency, anti-solidity, flexibility, softness and friendliness. Inflatability suggested the absence of permanence and the possibility of temporary environments which could be provided when and as required, and removed immediately afterwards. This suited the requirements of Pop as it emphasised the role of people rather than structures.

A number of designers set about finding ways of utilising this exciting new material. In England the visionary architectural group, Archigram, who in their own words sought 'flow and movement' and

Ron Herron. Capsule Pier. Archigram, England 1965.

Peter Cook. Hypothesis for 'Control and Choice' project. Archigram, England 1966.

'expandability and change' in the idealised urban projects that they were working on at this time, were among the first to see a role for inflatable structures. In a text accompanying their design of an interior for an exhibition held at the Woollands store in London in 1965 entitled 'The Breakthrough Designers', they claimed that 'chairs are on the way out and pneumatic seating, which forms to the body when sat on, will take its place'.[2] In the same year the architectural critic and historian, Reyner Banham, proposed his 'unhouse' which consisted of a plastic bubble,[3] and in 1967 Jean-Paul Jungman designed his experimental pneumatic house. The most dramatic inflatable structure actually built at this time was the Group Pavilion at Expo 1970 in Osaka.

While a number of architects played with the idea of blow-up buildings, the inflatable PVC chair became a widespread reality in the second half of the 1960s. The first example emerged, not surprisingly, from Italy, the country in which an 'Anti-Design' movement was most seriously proposed by a number of young architcts who sought an alternative to lush, formal, objects which had characterised Italian design in the late

1950s and early 1960s. The 'Blow' chair designed in 1967 by Paolo Lomazzi, Donato D'Urbino, and Jonathan De Pas, and manufactured by Zanotta – one of the more adventurous of the myriad of small furniture manufacturing companies located in and around Milan – provided a model which many others were to follow swiftly. In Britain a sequence of blow-up chairs came into the market-place in the following year among

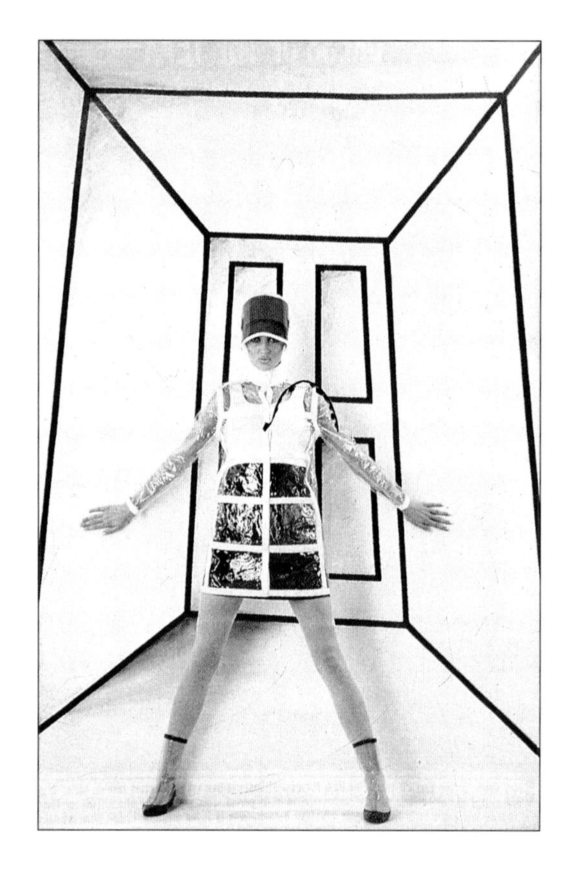

them the 'Pumpadinc' armchair, manufactured by Incadinc, available in white, red, navy-blue or purple, and an inflatable pouffe, designed by Arthur Quarmby and made by Pacamac in Lancashire. Quarmby's chair could be blown up with a cardboard balloon pump and both transparent and red versions of it were available. The Sunday colour supplements, by far the most receptive of the design-oriented magazines to the objects of Pop, illustrated them thereby making them available to a potentially large audience of consumers. Yet another British inflatable chair was designed by the Pop illustrator Roger Dean for the Hille company. Dean's version was covered in synthetic red fur thereby completing the destruction of the links between plastics and high culture.

Perhaps the most widely publicised inflatable PVC furniture, however, was designed by Quasar Khanh in France. His chairs and sofas were made up of a number of elements joined together with metal rings and could be filled with air, coloured gas or water.

The fashion for inflatable furniture was, characteristically of the objects of Pop, short-lived but it remains one of the most striking innovations of the late 1960s, suggesting a whole new way of using furniture in interior, or indeed exterior spaces. (Zanotta claimed that 'Blow' could be used as a means of reading a newspaper in a swimming pool.) They eroded, through their portability, the distinction between indoors and outdoors and eliminated the necessity of furniture being seen as 'fixtures'. PVC found another outlet within Pop culture, however, besides inflatable furniture. Its combination of strength with flexibility made it the ideal material for the fashion designers, inspired by the work of the Parisian Courreges, who sought to translate the imagery of science-fiction and space-travel into fashion items. Writing about the sci-fi look, which appeared suddenly in 1965, Meriel McCooey explained that 'in the last two years plastic has been much improved and scientists are

Plastic clothing by Michele Rosier. Paris 1966.

PVC dress. Denmark 1966.

Inflatable PVC 'Blow' chair, designed by Scolari, De Pas, D'Urbino and Lomazzi for Zanotta SpA. Italy 1967.

discovering it has as many permutations as the football pools',[4] added to which, in keeping with the 1960s designers' commitment to technology and the world of the future 'it's a material you can't work nostalgically'.[5] The pages of the fashion magazines from this year were filled with PVC halter-top dresses, space-shaped visors and models striking robot-like poses, all of them inspired by Courreges who dressed his models in white plastic glasses with narrow slits and calf-high white plastic boots with no toes and bows round the top – a direct attack on the middle-aged sophistication of haute couture as it had developed through the 1950s and early 1960s. With space travel becoming a reality by the mid 1960s, items such as silver anoraks closely

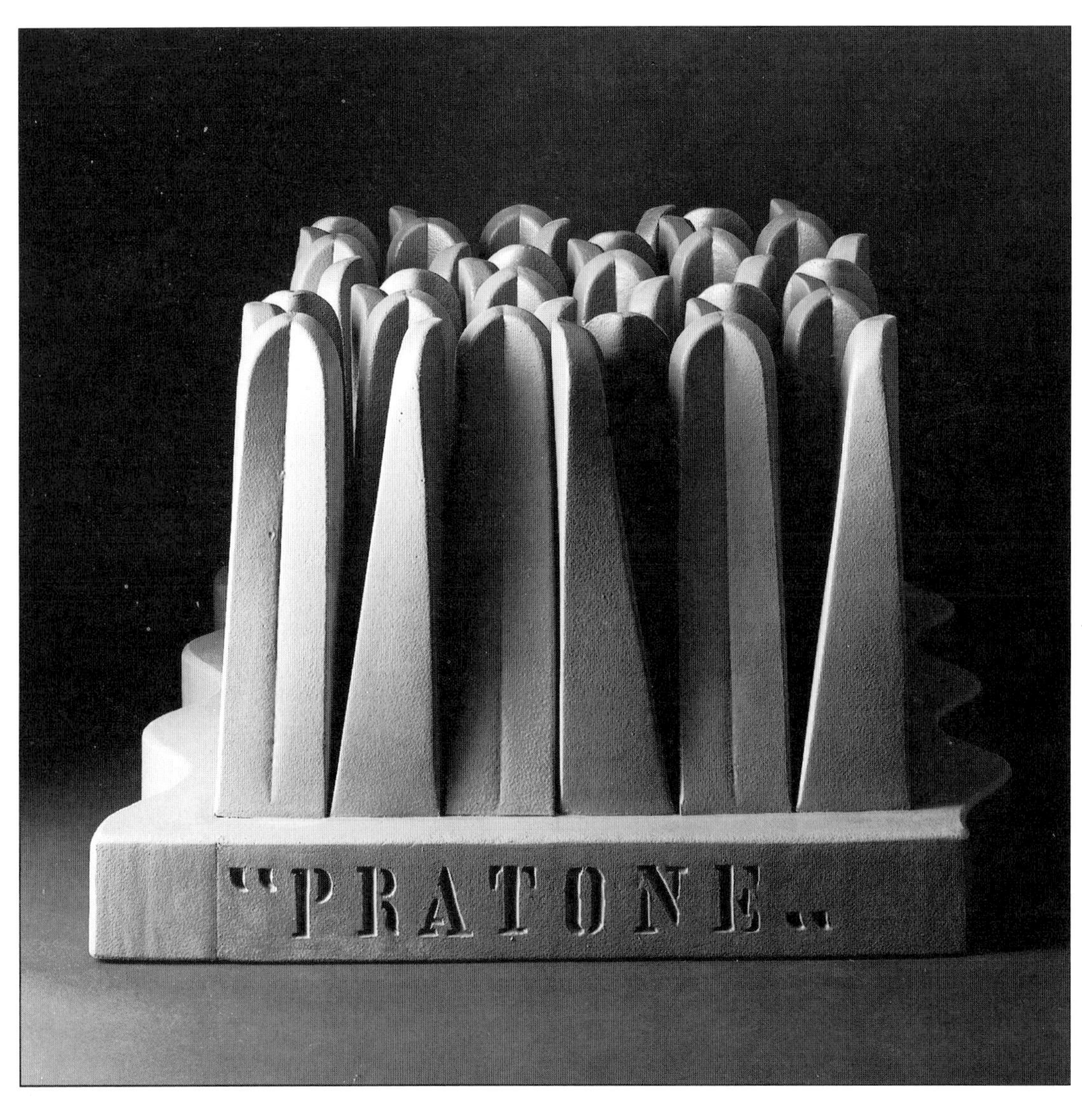

'Il Pratone' seat in polyurethane foam, designed by Gruppo Strum for Gufram. Italy 1971.

resembling astronauts' suits appeared. The break in the strict alliance between form and content which had occurred with Pop meant that anything could be turned into fashion, and space travel provided a highly evocative source of imagery which could be transferred as easily into a design for a dress as one for a kitchen. Where fashion was concerned the essential synthetic nature of plastics – predominantly PVC – symbolised modernity and availability and the appeal of its surface provided the sensorial instantaneity required for the objects of Pop.

The other plastic to achieve popularity in the second half of the 1960s – once again in the area of furniture – was polyurethane. Available, by this date, in varying degrees of rigidity it provided the ideal material for furniture which, while being both permanent and solid rather than disposable and immaterial suggested, nonetheless, more flexible ways of using seating items other than the conventional ones.

Once again Italy pointed the way forward and, under the banner of the 'Anti-Design' movement a number of items were produced

'I Sassi' seats in polyurethane foam, designed by Piero Gilardi for Gufram. Italy 1967.

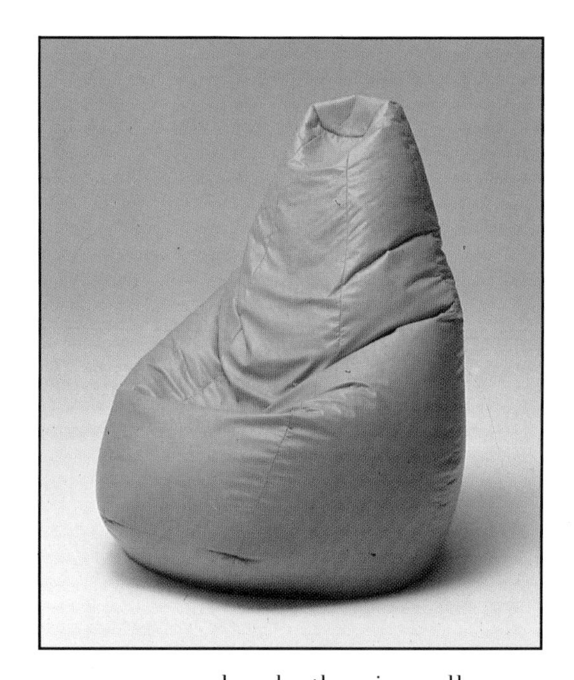

'Sacco' seat. Polyurethane pellets in a sack, designed by Gatti, Paolini and Teodoro for Zanotta SpA. Italy 1969.

– some mass-produced, others in small-scale, or prototype production only – which challenged the monumentality of mainstream modern Italian furniture. The radical architectural group, Archizoom – strongly influenced by Archigram – designed an organically-shaped sofa bed, named 'Superonda' (Superwave) in 1966 which exploited the sculptural potential of upholstered polyurethane foam, while the sculptor, Piero Gilardi, produced his 'Rocks' in the following year – lumps of painted polyurethane foam which looked like hard pebbles but in fact gave way when someone sat on them. In 1969 another of the radical Italian architectural groups to emerge in the late 1960s – Gruppo Strum – exploited the same ambiguous qualities of the material in their 'Big Meadow' seat, manufactured by Gufram. The use of polyurethane to play games with hardness and softness proved a vital strategy in the Italian movement. The idea of the 'disintegration of form' was one of the ways in which Pop served to break away from Modernism and open the doors to a new epoch of design in which form gave way to surface.

Other significant Italian contributions to Pop furniture included Gaetano Pesce's series of UP armchairs designed for C and B Italia. 'Up 2' of 1969, made of polyurethane and covered in stretch fabric, expanded from a flat pack on release from compression within its packing. The same sort of 'now you see it, now you don't' shock tactic which inflatables had utilised to such effect was here being reworked in another way. Finally, the famous and much emulated 'Sacco' seat, designed by Piero Gatti, Cesare Paolini, and Franco Teodoro, and manufactured by Zanotta in 1969, represented the ultimate example of flexible formlessness. Filled with thousands of small polyurethane pellets the seat adopted the shape dictated to it by its user, thereby rejecting completely the idea of absolute form and static monumentality. As did the inflatables, this flexible structure allowed the emphasis to shift from the 'sat upon' to the 'sitter', a crucial transformation in the ethics of the Pop revolution.

The Italian Anti-Design movement was strongly influenced by British Pop culture of the mid-1960s seeing there a model of mass culture asserting itself in the face of high culture. It was more directly committed, however, to the role of objects as metaphors of liberation and used them extensively as a mean of renewing the language of form. Italy's strong traditions of furniture-making and its earlier commitment to making lasting forms out of plastic combined to provide it with a strong manufacturing and design culture which could contain its own opposition. Thus the statements made were highly self-conscious and articulate.

One final Italian use of plastics as a material to contain 'alternative' values could be found in the work of the architect-designer Ettore Sottsass who, in 1966, exhibited a range of prototype wardrobes which were covered in sheets of plastic laminate, manufacture by ABET-Print. The use of these flat sheets provided him with a surface on to which he could apply the motifs and decorations that he culled from the Op and Pop paintings that he saw in the USA. This use of plastics as a neutral surface on to which 'meaningful' decoration could be applied aligned furniture with graphic and fashion design both of which made extensive use of the 'decorative surface' in the mid-1960s. It helped to complete the transition from object to image which was a vital contribution from Pop to cultural movements which came after it. In Britain the application of motifs such as flags and bulls' eyes to a wide range of Pop artefacts – trays to mugs to washing-up cloths – represented the pre-eminence of surface over form.

Ettore Sottsass. Wardrobes covered with plastic laminate. Poltronova. Italy 1966.

This theme was to re-emerge in the 1980s as a key aspect of what came to be called Post-Modernism. By this time it had also become apparent that plastics were materials lacking an inherent form or visual identity but suggesting an almost infinite range of possibilities, capable of carrying a number of different messages. Their propensity to change and metamorphosis is practically limitless and their chameleon-like character makes them eminently appropriate to a culture which thrives upon pluralism.

This idea was already evident in the 1960s within the context of Pop culture. The lasting lesson of the 1960s is that plastics could represent both mainstream and alternative culture without a hint of contradiction. The increasing range of plastic materials made available by the intensified research programmes of these pre-oil-crisis years meant that there was a plastic available for whatever cultural statement was needed. Both hard and soft, monochrome or decorated, transparent or opaque, plastics discouraged a reductive approach towards design and encouraged, instead, an expansion of possibilities. Thus the boundaries between 'high' and 'low' culture began to shift and the impact of the Pop revolution, helped along by the new plastics, changed things irrevocably and pointed the way forward to the pluralistic culture that characterises the last decades of the twentieth century.

Notes

1. R. Banham. 'Who is this Pop?' in *Motif*, 10, 1963, p.12
2. P. Cook. Quoted in R. Banham. 'A Clip-On Architecture' in *Design Quarterly*, June 1965, p.3
3. R. Banham. 'A House is not a Home' in *Art in America*, April 1965
4. M. McCooey. 'Plastic Bombs' in *Sunday Times Colour Supplement*', 15 August 1965, p.21
5. Ibid.

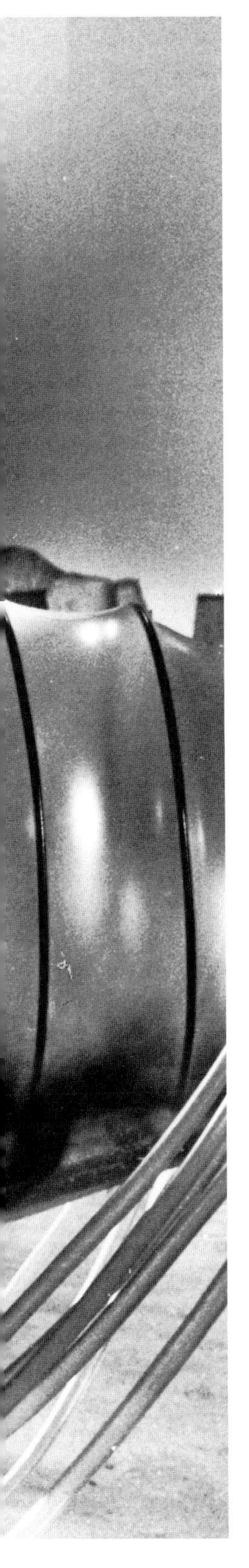

Jane Fonda in see-through membrane; still from *Barbarella*. 1968.

TRIUMPH OF SOFTWARE

Reyner Banham

If we *needed* the concept of a fur-lined spaceship (and we did, even if we didn't know it), we have it now. We also have a female astronaut stripping off in free fall, beginning very conventionally with the gloves like any old gipsy rose. And behind her, in the fur-lined control module, there are a rococo nude in plaster and an airlock door panelled with a section of Seurat's *Dimanche à la Grande Jatte*. And all this before the credit titles are properly out of the way, to sum up what *Barbarella* is going to be all about for its devotees; thus:

Item: an unexceptionally straight-up-and-down view of sex – Jane Fonda's clean little surf-girl face (Nancy Sinatra might have been even better in the part!) and Terry Southern's sanitized *Candy*-coloured script combine to produce a love-object about as kinky as the All England Lawn Tennis and Croquet Club.

Item: a preoccupation with a kind of environment that, if not always fur-lined, is always with-it – so much so that it will become what the film is remembered to have been about in the rather specialized circles where it is remembered.

And item: a trickle of visual jokes – art-jokes, rather – that need an audience of at least DipAD complementary studies level. Since the movies are the one visual art-form whose criticism we entrust to the visually ignorant, all this art-jokery completely missed the assembled body critical at the press show, and I would like to congratulate the four people who laughed in the same places as I did, not one of them a major weekly or daily critic as far as could be seen.

Predictably, therefore, *The Guardian* doubted if it would ever become a cult movie. I have news for you, mate: old, cold news. *Barbarella* has been a cult movie for months, ever since the first stills were in *Playboy* and the colour supps, and were pinned up in architecture student pads and studios across the world. For *Barbarella*, unlike most feature movies, which are about two years behind the visual times for unavoidable mechanical reasons, is barely twelve months of of date. Style-wise, it falls about halfway between the Million Volt Light-Sound Raves at the London Roundhouse at the beginning of last year, and the great Arthur Brown/Jools/Inflatable rave organised by Architectural Association students there at the end of the year.

Compare Stanley Kubrick's *2001*, since everybody does. My spouse's verdict that this film's psychedelic (eh?) sequences 'would have been great if they had been at Hornsey two years ago' marks the point at which Kubrick most nearly caught up with the live visual culture of now. The rest of the movie was like Pompeii re-excavated, the kind of stuff that Richard Hamilton had in his *Man, Machine and Motion* exhibition back in 1955. All that grey plastic and crackle-finish metal, and knobs and switches, all that . . . yech . . . *hardware*!

But *Barbarella* is the first post-hardware SF movie of any consequence. By one of those splendid coincidences that used to make German historians believe in the Zeitgeist (and which English historians always miss) the film was premiered here in the same week that a company called Responsive Environments Corporation went public on the New York stock exchange. Whatever the company is about *Barbarella* is about responsive environments, of one sort or another, and so has been the architectural underground for the last three years or so; and from where I stand, I can't see how this could avoid becoming a cult movie. . Responsive environments in the sense of not being rigid and unyielding; articulated only by hinges between disparate rigid parts: an ambience of curved, pliable, continuous, breathing, adaptable surfaces.

Fur is exactly such a surface. Of all the 'materials friendly to man' it is the friendliest, because it is kissing cousin to our own surface and grows in some of our friendliest places. But is also has, in the most objective and quantifiable terms, physical properties that would make it worth inventing if it did not exist: flexible, shock-absorbing, heat-insulating, acoustically absorbent and selectively responsive to reflecting light. (Also smelly and difficult to keep clean? So are you, *hypocrite lecteur*!)

It is extremely difficult to think of a better material to line a spaceship or any other vehicle in which human flesh is going to be tumbled about during magnetic storms, free fall or rough re-entries. Or tumbled about in that grand old Shakespearian sense of the word; Barbarella gets it once in/on furs, and again in an angel's nest lined with what has to be moss, the vegetable kingdom's nearest equivalent.

Both fur and moss, however, exist already; no point inventing them. What has been invented, and recently enough to have been almost totally overlooked by the movies, is the inflated or otherwise distended or tension transparent plastic membrane; and this is *Barbarella*'s other great environmental hang-up. Rightly so, because inflatables too are . . . yes, friendly. I actually heard that precise word applied during an evening illuminated balloon-happening in Los Angeles. The balloons were the army-surplus sort that get advertised in SF magazines, around eight feet in diameter, blown to a merely floppy condition with an air/helium mixture that made them only marginally buoyant.

Trying to hold them steady against the light night-wind was a bit like trying to handle a plump, drunk, amiable, unstable girl at a party. Suddenly you'd find you were half-smothered in bulging, yielding, demanding plastic that wouldn't go away when you pushed it. And there was this bird came up (I think she'd been sniffing the helium) and said could she borrow a balloon because they seemed kinda, well, you know, friendly; vanished with it into the professionally designed landscaping; and reappeared about 20 minutes later, weeping and freaked out, with its exploded remnants.

Barbarella digs these (and other) aspects of inflatables in depth and at length. She sleeps (lit and photographed from below) on a transparent membrane that dimples to her form. The sails of the ice yacht become erectile when the wind blows, and the fur-trimmed tumble takes place in the yacht's translucent 'tail'. In the wicked city of Sogo, inflatable bolsters bumble loosely about the interiors; bodies are trapped, or rescued, through transparent flexible plastic tubes; the Black Queen manifests herself out

Still from *2001*. 1968.

of an exploding plastic bubble, and her dream-chamber is a bubble furnished with smaller bubbles and giant thistledown (vegetable fur again) from which she and Barbarella escape during the final Gotterdamaround in a bubble of innocence.

The whole vision is – significantly, I suspect – one in which hardware is fallible, and software (animate or otherwise) usually wins. Barbarella's spaceship is more often broken down than not. The electronic gadgetry in David Hemmings's revolutionary HQ always goes on the blink when he needs it. Milo O'Shea's positronic ray machine fails to make him master of the city. More to the point, his Excessive Machine fails to kill Jane Fonda with pleasure – instead, the insatiability of her flesh burns its wiring and blows its circuits, thus giving O'Shea the best line in the script: 'Have you no shame?'

All along, brittle hardware is beaten by pliable software. Barbarella out-yields the

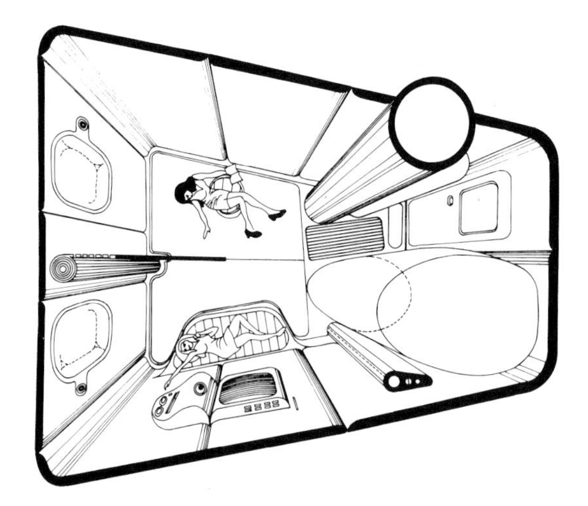

excessive Machine, and the monumental structure of the city of Sogo finally falls to subterranean software: the dreaded Mathmos. Roughly speaking, the Mathmos is (cut-price French metaphysics aside) a blend of a Mark Boyle/UFO light show, detergent commercials, and a circa-1940 pulp novel trying to describe orgasm (you know: iridescent liquid fire seemed to surge up through her etcetera and like that). If you could get it in cans, it would be a great product. You could pour it on the carpet and have a psychedelic wade-in.

In the city of Sogo, however, it glints gurkily around in paramoebic blobs under the glass floors, doubling the roles of a communal libido-sewer and the London Electricity Board. 'It feeds upon our evil, and in return gives us heat, light and power.'

It has to be kept under glass, literally, because otherwise it will destroy the city. This is a conventional type of SF situation, too corny in itself to be worth commenting; but in the context of *Barbarella*, it seems to reveal something of the inner tensions between the logic of the Now and the restraints of past tradition that SF shares with all other human activities, in spite of its preoccupation with futures.

At first sight, the city of Sogo is ridiculous. It doesn't look like Sodom, Sybaris or Las Vegas, which is understandable. But neither does it look like anything that could grow naturally out of the inflatable/fur-lined technology that dominates the rest of the movie, and the software culture within it. Instead, the city looks – both without and within – uncommonly like an architectural student megastructure of the post-*Archigram* age: multi-storey frames carrying nodes and capsules of living-space above ground, a labyrinth of tunnels below.

Warren Chalk. 'Capsule Homes Interior'. Archigram. England 1966.

But historically this is reasonable. Both *Barbarella* in its original French cartoon-strip form, and *Archigram*'s plug-in city project are half-jokey European intellectual derivatives from basic US pulp SF. Both stand in a tradition of environmental visions that runs back visually through the comics, and verbally through texts like Isaac Asimov's *Caves of Steel* (now in Panther paperback; don't just sit there, go and buy it!), to Futurist architectural visions of the nineteen-teens and ultimately to basic sources like H. G. Wells – notably *The Sleeper Awakes*, which is still a sacred text among architects.

This tradition conserves an essentially nineteenth-century vision of the urban environment – densely built, over-populated, low on privacy, violent, serviced by public transport. If that sounds just like New York, it is also a city type whose built form would make Colin Buchanan feel at home, the kind of city that most architects would prefer to Los Angeles, the city seen (like Cumbernauld town centre) as a single artefact, 'le plus grand outil de l'homme' (Le Corbusier: how did you guess?). Asimov had the vision and craft to construct an alternative city in another book of his, *The Naked Sun*; but it is a pretty rare thing in SF. *Barbarella* holds determinedly to the grand old Wellsian tradition, and Sogo is seen, in distant views, as a singular construction on top of a hill.

If seen as a piece of hardware, that is. And just as present technology and culture make it possible and necessary to construct alternatives to the high-density architect-preferred city, so the extrapolation of present culture and technology, along the vectors implicit in *Barbarella* (and *Archigram* too), makes the survival of the Wellsian artefact-city inconceivable. *Archigram* has acknowledged the logic of its situation by progressively abandoning its megacity visions in favour of ever more compact, adaptable and self-contained living capsules which, by last summer, had shrunk to the proportions of a rather complex suit which could be inflated (for real; the prototype worked) to provide everybody with their own habitable bubble of innocence.

The dramatic structure (for want of a better phrase) of *Barbarella* does not lend itself to such gradualism, nor its internal logic (eh?) to such commonsense solutions. It is intolerable that a lump of hardware like the city of Sogo could co-exist with the living, breathing vision of a friendly, sexy, adaptable personal environment. The Black Queen pulls the plug on the Mathmos and the whole hardware scene goes down in boiling flames.

Milo O'Shea presides over this final solution at the console of his positronic ray machine, screaming defiance at his fate with a grand raving Wagnerian glee. He reminded me, somewhere deep down inside, of one or two characters who might be president of the International Union of Architects in about a generation . . .

'Cushicle' exhibited at the Milan Triennale. Italy 1968.

PLASTIC

Roland Barthes

Despite having names of Greek shepherds (Polystyrene, Polyvinyl, Polyethylene), plastic, the products of which have just been gathered in an exhibition, is in essence the stuff of alchemy. At the entrance of the stand, the public waits in a long queue in order to witness the accomplishment of the magical operation par excellence: the transmutation of matter. An ideally-shaped machine, tabulated and oblong (a shape well suited to suggest the secret of an itinerary) effortlessly draws, out of a heap of greenish crystals, shiny and fluted dressing-room tidies. At one end, raw, telluric matter, at the other, the finished, human object; and between these two extremes, nothing; nothing but a transit, hardly watched over by an attendant in a cloth cap, half-god, half-robot.

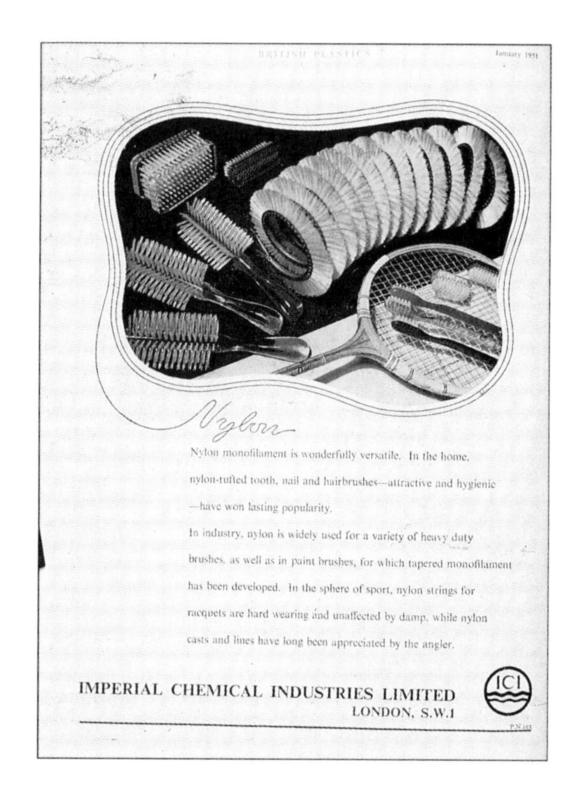

An advertisement for ICI, showing a range of nylon products. *British Plastics*, Jan 1951.

So, more than a substance, plastic is the very idea of its infinite transformation; as its everyday name indicates, it is ubiquity made visible. And it is this, in fact, which makes it a miraculous substance: a miracle is always a sudden transformation of nature. Plastic remains impregnated throughout with this wonder: it is less a thing than the trace of a movement.

And as the movement here is almost infinite, transforming the original crystals into a multitude of more and more startling objects, plastic is, all told, a spectacle to be deciphered: the very spectacle of its end-products. At the sight of each terminal form (suitcase, brush, car-body, toy, fabric, tube, basin or paper), the mind does not cease from considering the original matter as an enigma. This is because the quick-change artistry of plastic is absolute: it can become buckets as well as jewels. Hence a perpetual amazement, the reverie of man at the sight of the proliferating forms of matter, and the connections he detects between the singular of the origin and the plural of the effects. And this amazement is a pleasurable one, since the scope of the transformations gives man the measure of his power, and since the very itinerary of plastic gives him the euphoria of prestigious free-wheeling through Nature.

But the price to be paid for this success is that plastic, sublimated as movement, hardly exists as substance. Its reality is a negative one: neither hard nor deep, it must be content with a 'substantial' attribute which is neutral in spite of its utilitarian advantages: *resistance*, a state which merely means an absence of yielding. In the hierarchy of the major poetic substances, it figures as a disgraced material, lost between the effusiveness of rubber and the flat hardness of metal; it embodies none of the genuine produce of the mineral world: foam, fibres, strata. It is a 'shaped' substance: whatever its final state, plastic keeps a flocculent appearance, something opaque, creamy and curdled, something powerless ever to achieve the triumphant smoothness of Nature. But what best reveals it for what it is is the sound it gives, at once hollow and flat; its noise is its undoing, as are its colours, for it seems capable of retaining only the most chemical-looking ones. Of yellow, red and green, it keeps only the aggressive quality, and uses them as mere names, being able to display only concepts of colour.

The fashion for plastic highlights an evolution in the myth of 'imitation' materials. It is well known that their use is historically bourgeois in origin (the first vestimentary postiches date back to the rise of capitalism). But until now imitation materials have always indicated pretension, they belonged to the world of appearances, not to that of actual use; they aimed at reproducing cheaply the rarest substances, diamonds, silk, feathers, furs, silver, all the luxurious brilliance of the world. Plastic has climbed down, it is a household material. It is the first magical substance which consents to be prosaic. But it is precisely because this prosaic character is a triumphant reason for its existence: for the first time, artifice aims at something common, not rare. And as an immediate consequence, the age-old function of nature is modified: it is no longer the Idea, the pure Substance to be regained or imitated: an artifical Matter, more bountiful than all the natural deposits, is about to replace her, and to determine the very invention of forms. A luxurious object is still of this earth, it still recalls, albeit in a precious mode, its mineral or animal origin, the natural theme of which it is but one actualization. Plastic is wholly swallowed up in the fact of being used: ultimately, objects will be invented for the sole pleasure of using them. The hierarchy of substances is abolished: a single one replaces them all: the whole world *can* be plasticized, and even life itself since, we are told, they are beginning to make plastic aortas.

Polythene cups and food containers. Tupperware Plastics Company. USA 1940s-1960s.

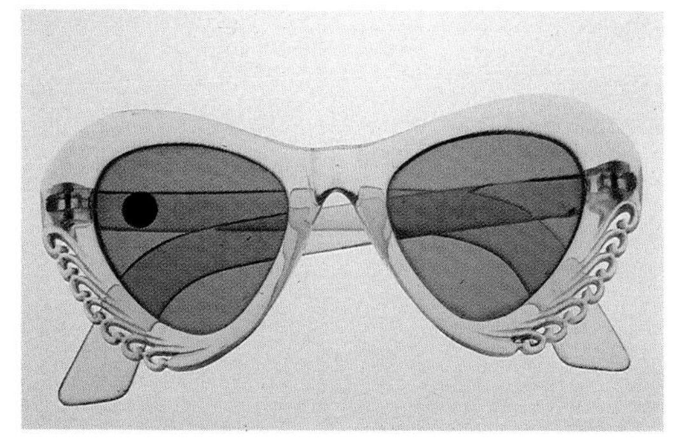

Plastic sunglasses. 1950s.

NATURAL WOOD, CULTURAL WOOD

Jean Baudrillard

. . . The same thing goes for materials as well. Take wood for example, which is in such demand today because of its emotional associations: It takes its substance from the earth, it seems to live, to breathe and 'to work'. It has an intrinsic warmth; it doesn't reflect things like glass but seems to burn from within; it holds time in its fibres, it's the ideal receptacle because one wants to make sure that everything contained within it escapes the progress of time. Wood has its own smell, it ages, it even has its own parasites etc. In short this material is a living being. The image of the 'richness of oak' which all of us have in our mind's eye, evoking successive generations, heavy furniture, and family homes, is like this. But, does the 'warmth' of this wood, as well as that of freestone, of natural leather, of woven fabric, of beaten copper etc. (all these elements of a material and maternal dream which today nourish our nostalgia for the days of luxury) retain its meaning today?

In the contemporary environment, all the organic or natural materials have, in practice, found their functional equivalents in plastic or compound substances: wool, cotton, silk and linen have found their universal substitutes in nylon and its countless variations. Wood, stone and metal have relinquished their position to concrete, Formica and polystyrene. There's no possibility of denying this evolution or of dreaming idealistically of the warm, human substances of yesterday's objects. The opposition between natural substances and synthetic substances, like the one between traditional colours and those used today, is only a moral opposition. Objectively speaking, materials are what they are: there are no true or false ones, natural or artificial ones. Why should concrete be thought of as less 'authentic' than stone? We experience old synthetic materials, such as paper, as completely natural, and glass is one of the richest materials in existence. In essence, the inherited nobility of materials forms part of a cultural ideology which is analogous to that of the aristocratic myth in the human domain, and even this cultural prejudice evaporates with time.

The important thing is to see, outside the vast perspectives that these new substances have opened up for the world of practice, the way in which they have modified our sensorial relationship with materials.

In the same way as, where colours are concerned, the tonal spectrum (warm or cold or intermediary) signifies a moving away from their moral and symbolic status towards an abstraction which makes possible systematisation and games, so the manufacture of synthetics signifies for materials a stepping back from their natural symbolism towards a polymorphism, towards a superior level of abstraction which enables a game of the universal association of materials to take place, and therefore a move

Ideal Home
April, 1955
"Whenever I think of surfaces......
I think of FORMICA"
LAMINATED PLASTIC REGD.
Blinds by Sunway
This housewife is a 'FORMICA' kitchen enthusiast—but aren't we all? Won't *you* feel life is good when you own a kitchen where all the surfaces are jewel-bright clean-at-a-wipe 'FORMICA' Laminated Plastic? You can buy your dream-kitchen unit by unit or complete, or you can fit 'FORMICA' panels to existing surfaces. 'FORMICA' material will not crack or chip when fixed, and will not stain in normal use. And it comes in a rainbow-choice of gay 'Linette' and Softglow patterns.
N.B. Whether you buy 'FORMICA' Laminated Plastic on kitchen units, other furniture or in do-it-yourself panels, ask for it by name—pronounced 'FORMICA' as in 'for MY kitchen'. Look for this stamp on every piece: it washes off *easily* with soap and water.
FORMICA LAMINATED PLASTIC MADE BY DE LA RUE
the surface with a smile
FREE LEAFLETS *from Thomas De La Rue & Co. Limited (Plastics Division) Dept. E.12, 84-86 Regent St., London, W.1. Tele: REGent 2901*
'FORMICA' *is a registered trade mark and Thomas De La Rue & Company Limited is the registered user.*
126

An advertisement for Formica Limited. *Ideal Home Magazine.* *April 1953.*

beyond the formal opposition of natural materials and artificial materials: there is no longer, today, a 'natural' distinction between a glass partition wall and wood, rough concrete and leather: 'warm' values or 'cold' values, they all come under the same heading of material-elements. These materials which in themselves are disparate are homogeneous as cultural signs and can be integrated into a coherent system. Their abstraction allows them to be combined.

OBJECTS AND THEIR SKIN

Ezio Manzini

Once upon a time the world was dominated by the looming outlines of locomotives, ocean liners, steel bridges, great massive constructions – a universe peopled by shapes that filled space with their three-dimensional solidity, shapes that were obstructively physical, whose 'truth' lay in their structure, their working, in the intrinsic quality of the material of which they were built.

In that world, the word 'surface' often held overtones of 'superficiality' – something not absolutely essential, something rather frivolous if not downright improper.

Michele de Lucchi. 'Kristall' side-table. Memphis, Italy 1981.

But that world which we have just been recalling, born in the Industrial Revolution two hundred years ago and coming of cultural age with the Modern Movement at the beginning of the present century, just doesn't exist anymore. Or rather what remains of it is only the objects that went to make it up, and which have now been subordinated to a new technical and cultural system that modifies the priorities with which we view their performance, that changes the criteria according to which they are appreciated or otherwise, and revolutionises the mental models with which one interprets – or tries to interpret – the new artificial environment.

In this new environment, the surface is no longer constrained to pretend that it just does not exist, and almost as if in vindication of the obscurity to which it was relegated for so long it is now taking on a new dramatic role that upstages the third dimension.

Photo-luminescent laminates from the 'Lumiphos' range. Abet Laminati. Italy 1980s.

Indeed, the emblematical images of the present-day world reveal an environment tendentially dematerialised, as fluid as the flow of information that passes across and through it, flattened down into the two-dimensionality of printed paper and the television screen.

And that's not all. This prevalence of the two-dimensional (and the dematerialisation that it implies) seems to go much further than just the boundaries of the world of information and information science. Physical objects seem to have come under the same influence, as if by some strange drag effect; it's not just that the numerous family of objects that have been transformed by electronics and miniaturisation in the normal course of their evolution, but also those objects that by necessity and by their nature keep to their three-dimensional character that are now entrusting a greater part of their expressive capacity and their performance to the surface area. In view of all this, it is opportune at this juncture to take an overall look at surfaces and the role they play in the definition of the artificial environment.

Indeed, the question has already cropped up within the debate and praxis of design. The return of decoration has been one of the most meaningful and most widely talked of phenomena of the last ten years, and can be taken as a significant and evident aspect of the new prominence of the surface. But that is not all there is to the story. It ranges far wider and far further afield than the boundaries set by the vagaries of taste in decoration and arguments about the stylistic dogmas of the Modern Movement.

From Confines of Material to Interface

There's a very simple way of considering the surface of an object. It's the point where the material of which the object is made finishes and the outside world begins. But that definition is only acceptable in certain cases (a stone, for instance, or a plastic article that comes straight from the mould in its finished state, requiring no further treatment). And even in such cases as these the geometrical definition hides a much more complex physical reality, for the surface is, in fact, an outer layer of atoms and molecules that is called upon to stand up to demands and conditions far different from those that affect the inner layers. This 'front line' layer has to withstand all sorts of physical, biological and chemical stresses and strains.

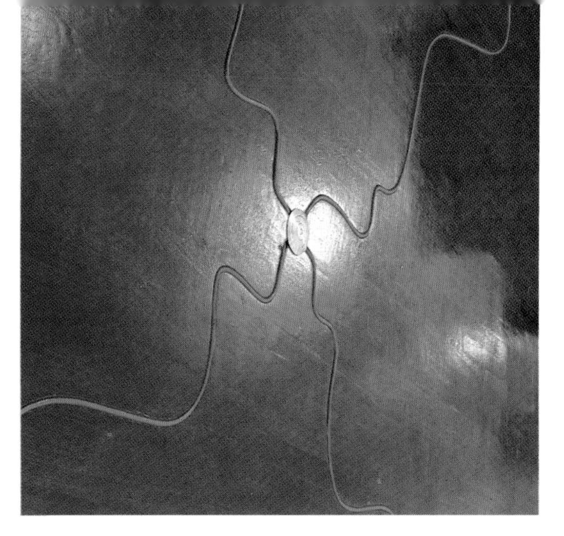

Detail from the Cassina showroom interior in Milan designed by Clino Castelli. Italy 1985.

Dome covered with a retro-reflecting material producing delicate lighting effects, designed by Clino Castelli for Cassina. Italy 1985.

Furthermore, it is the part of the object that determines the user/object relationship, so sensorial qualities (the look, the feel, the temperature) and symbolic and cultural values all come into it.

Now, looking at it from the opposite angle, from inside out as it were, we can add that if the surface of an object is the last layer of a material which continues with the same properties towards the inside, then it is a waste. Nature is thrifty, and it's no mere chance that all the more complex organisms have some kind of skin – a special kind of structure to keep the inside in and the outside out. And most manufactured articles, too, are subjected to some kind of surface treatment to make this outer layer more resistant and to improve its performance. So another definition is needed, a more comprehensive one, starting off with a reconsideration of the role of the surface and stressing its relatively autonomous character (in relation to the rest of the object, that is) and its dynamic qualities. Instead of a mute and static limit of the material we now have an interface between two environments whose role involves the exchange of energy and information between the two. The surface as an 'osmotic membrane' capable of favouring or obstructing this exchange process becomes itself a component of the object (a component that we may think of as two-

dimensional) that can mediate between the exterior and the interior of the object itself, or produce a range of performance autonomously.

Given these terms, thinking of the surface simply as that place where the object stops being is only one of very many different possibilities. The range of activities and performances that the surface can carry out is quite wide, and in a state of constant expansion. There are more obvious and traditional ones (it gives protection, and aesthetic and sensorial qualities to the material beneath), and those that transform

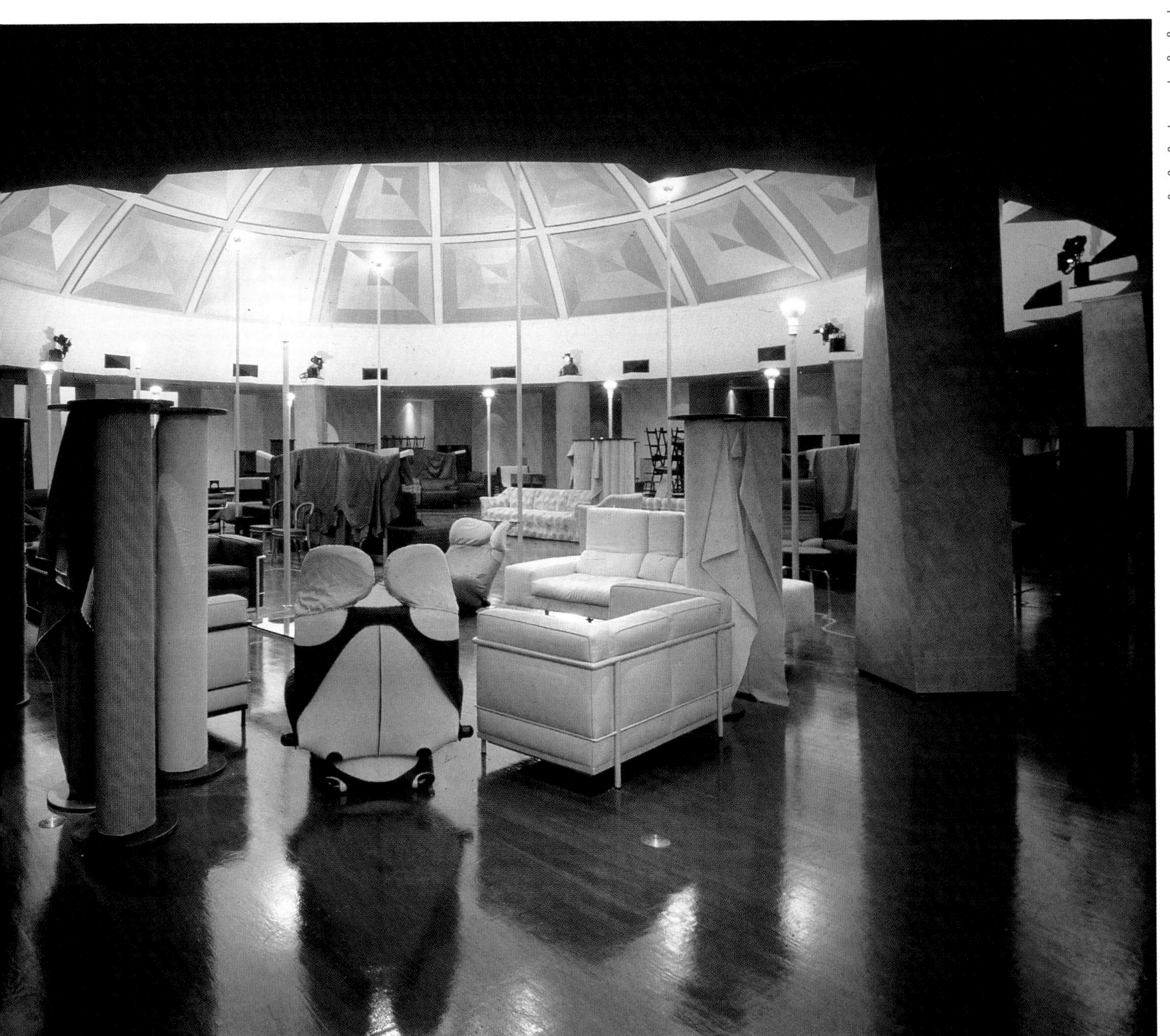

it into a support for static communication (printing on paper), and dynamic ones (surfaces sensitized with bidimensional input and output information components).

Ettor Sottsass. 'Veneziana' plastic laminate. Abet Laminati. Italy 1980.

Technical Background and Cultural Attitudes

The first artefacts that man produced were pieces of homogeneous material, picked up and treated in the simplest way to make the most of their intrinsic mechanical capabilities: stones, sticks, bones and so forth. In this very primitive technical phase, the surface represented only the limit to which the material extended, it is true, but it also provided an area for graphic symbols. The very first graphic representations known to us are regular incisions, notches, on sticks or pieces of bone. We don't known precisely what they were intended to mean, these scratchings perpetrated thirty thousand years ago, but they were surely repetitive and related, forming a rhythm. It is the same rhythm that in our present age forms one of the particular traits of that figurative expression that we call 'decoration'. Objects, writes Leroi-Gourhan, have been 'speaking' ever since paleolithic times, using their signs to tell of the reality, the characters, the cultural references of an ethnic group in some particular phase of its history. In those ages marked by a very thin layer of artifice, when there was limited technical control over the material and therefore over the structure of the objects made from it as well, the surface was a place upon which to express one's own image and identity. And so from the objects to the walls and wall paintings, and from the walls to the body and tattooing and body-painting.

Maybe if we study the recent developments of this long pictorial and representational saga, we may come to conclusion that the Modern Movement, with its declared aversion to decoration and surface ornament, was just stressing the element of novelty in the new technologies of the nineteenth century – a hitherto undreamed-of capacity of controlling the surfaces of manufactured objects. The invalidation of the surface and of the messages that it might bear stemmed from the wish to do away with anything that might detract from the geometrical purity of the volumes and shapes that could at last be produced – and reproduced – by mechanical means. The surface – the skin – the coating – call it what you will – clouded the vision of their effect perfect and essentially functional geometry.

Thereafter the culture of design rediscovered the value of the surface, that outer.layer, and of the sensorial variables that are built up upon it. Setting the abstract idea of the formal quality against the concrete and physical one of the sensorial quality, it first of all reappraised decoration for its own sake – which the Modern Movement had denounced as immoral – and went on to deal with the soft qualities of the object in hand; not only the visual qualities, but the tactile, thermal, and odiferous ones too.

But this rich facet procedure would be

pretty meaningless were it not for the contemporary technical transformations that were building up the background. The development of composite materials in which every layer has its own particular specialized function creates the problem – in the design area – of what particular quality the outermost layer should have, the skin, making nonsense of the 'frank image of the material' concept in the sense that the Modern Movement understood it; materials have a 'skin' and their image is that of the skin, with the range of variations that it can provide.

In other words, if the material can be projected, if controlled anisotopic and non-homogeneous properties can be produced, then the last layer – the surface – is no longer linked to the properties of the underlying ones but has a degree of autonomy. So physical and technological limits in the definition of the final image of the product tend to thin out, giving the designer a wide choice of surface finishes. In such conditions, the final choice is less and less justified by technical demands and those deriving from consolidated identity intrinsic to the material, and the surface of the finished product tends to turn out as a kind of screen upon which to put signs and references to a vast store of cultural material.

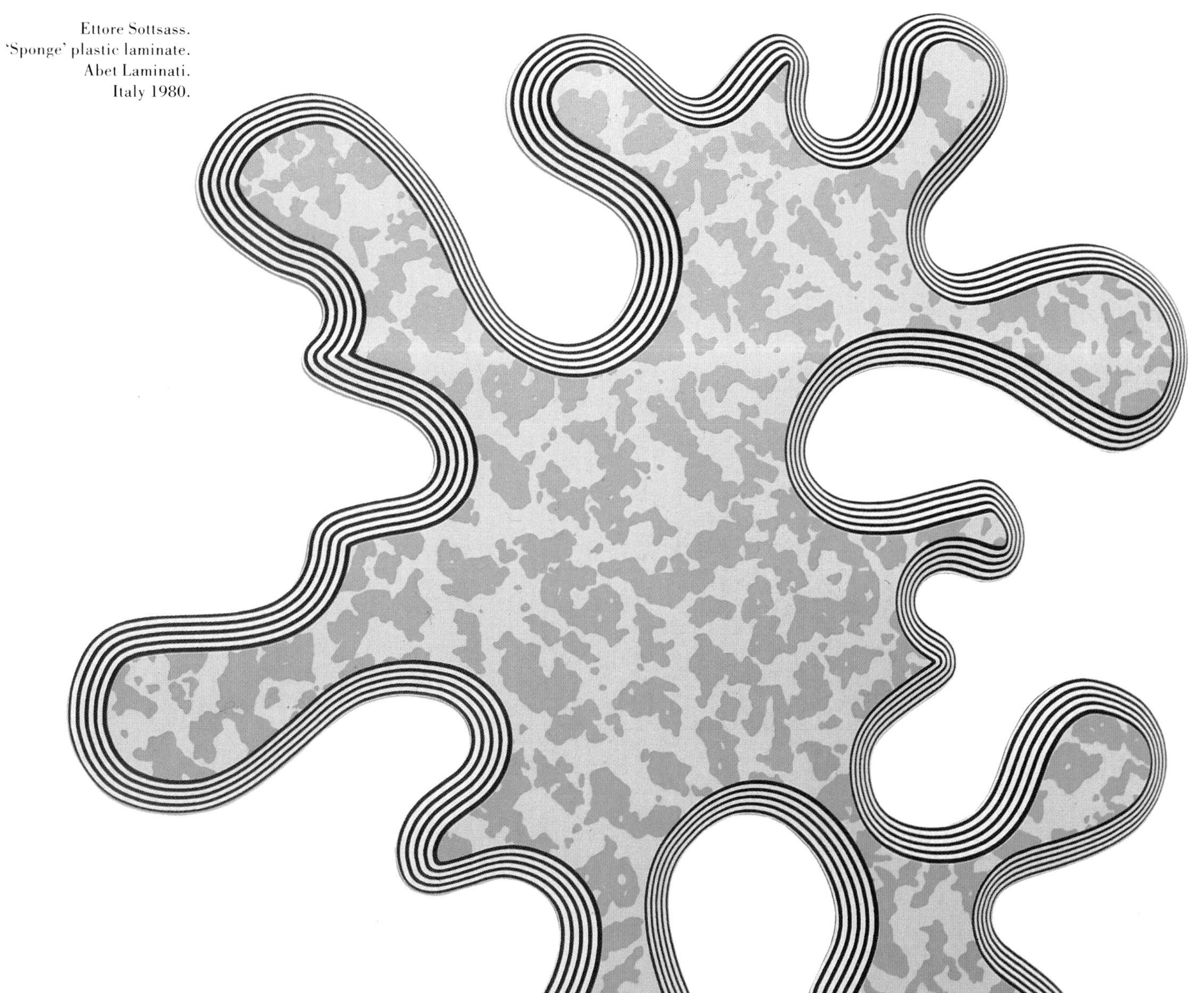

Ettore Sottsass. 'Sponge' plastic laminate. Abet Laminati. Italy 1980.

This new technical design atmosphere is reflected, too, in the different way that the objects produced within its confines are perceived, understood, and recognized. In the past, certain sensorial qualities would cause a surface to be definitively associated with this or that material, and that material – whatever it happened to be – with a set of technical and cultural values with certain associations contributing to the cultural standing of the object itself. But this is not so any longer.

The impossibility of knowing just which of all the infinite combinations of different materials, made possible and available by modern technology, lies beneath the surface of the article we are looking at leads to a kind of blunting of the image. The surface gives no hint at anything beyond itself; that is to say it presents the screen that has already been mentioned, upon which someone had projected signs and references.

Surface and Sensoriality

It is the surface quality of an object that gives us, through our senses, a knowledge of its shape and form. This particular process – the process which, for instance, transmits to our brains the information that a certain object is, say, square – is exclusively a human skill, but it is not the only nor even the surest way we have at our disposal of recognizing an object. The richness of our spatial-temporal experience makes it much more profound than one deriving from a relationship with geometric or functional abstractions.

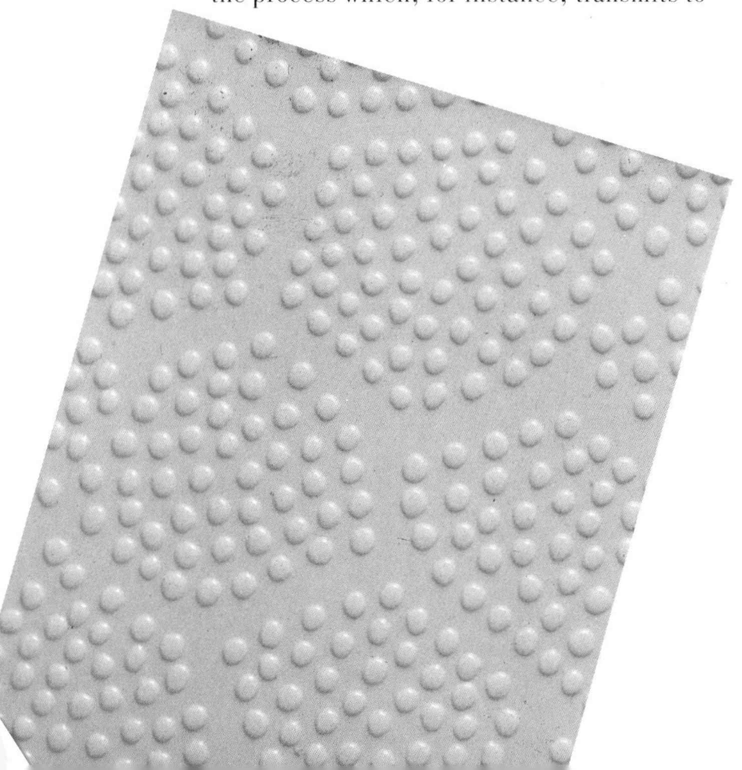

The recognition of a colour, or of a consistency, or of a texture (like that of a smell or of a taste) involves sensoral activities that are quite different from those called into play for the recognition of a shape. They demand a lower level of interpretation process, being concerned with more direct and closer physical contact. Colour awareness is the decodification of a wave-length, and tactile consistence that of a mechanical action. The world of the senses is an analytical one; synthesis of the image follows at a later stage and not necessarily even always then. Indeed, a part of this sensorial activity remains infra-symbolic, and beyond the reach of language, way down in the depths of our zoological being.

Undervaluing the importance of the surface, (and by that we mean the theatre of this intricate communication process), in order to stress the purity of the shape is, then, a particular aesthetic choice and not the general rule; a world of significant shapes and forms is all well and good, but, if that is all, homogeneous and uninteresting everyday surfaces lack a whole layer of sensorial relationships. On the other hand, although in the past the predominance of the natural environment provided an endless pattern-book of different surfaces, (and artificial and man-made substances were so little used that to choose one meant an

Left and opposite:
Range of decorative translucent laminates. Abet Laminati. Italy 1980s.

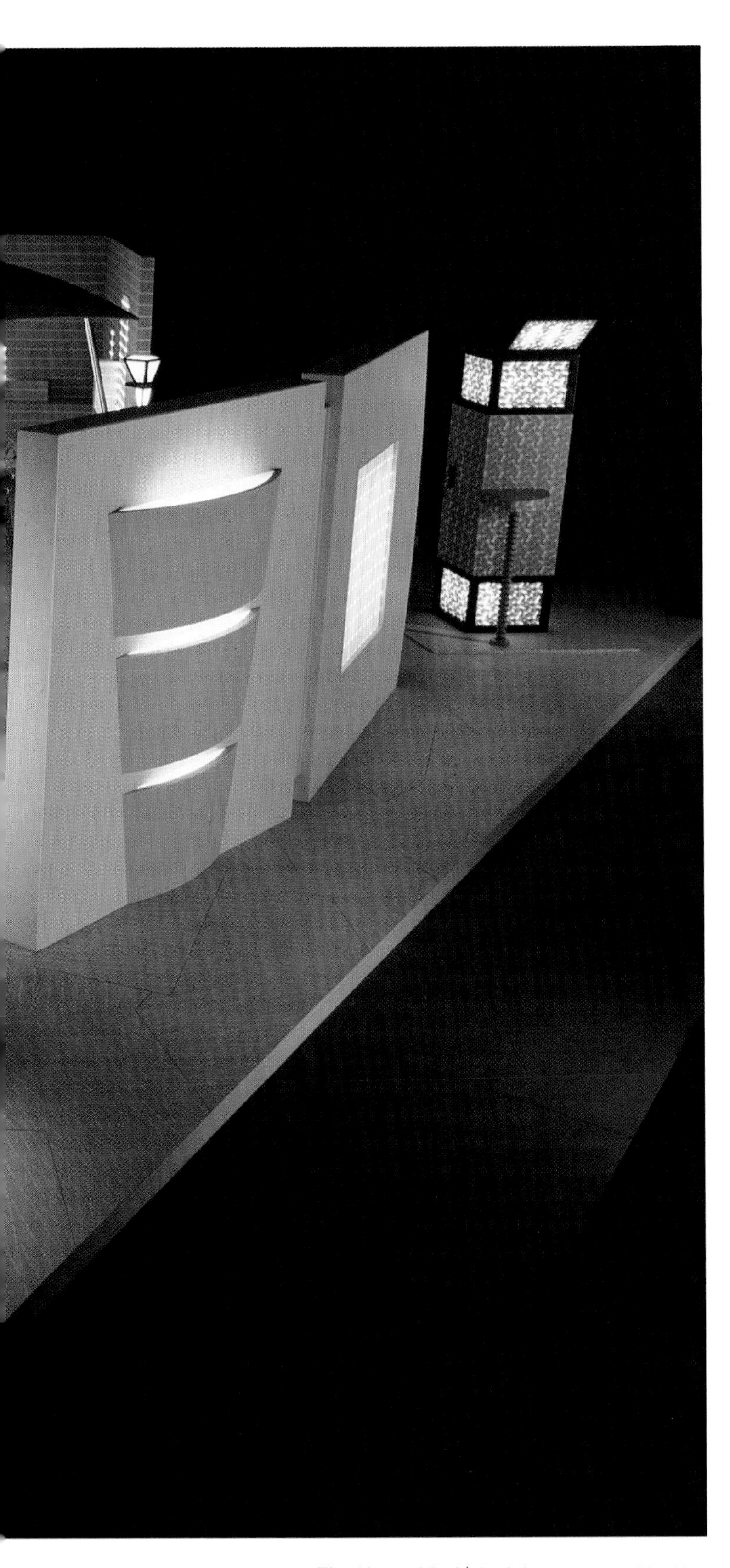

The 'Material Lights' exhibition, curated by Abet Laminati, featuring plastic laminates designed by a group of young Italian designers and manufactured by Abet Laminati. Italy 1987.

immediate and automatic down-grading of the quality of the finished product), nowadays the number of artificial substances in general use has grown by astronomic proportions, so that the variety of the surfaces is by now an inbuilt element of the design, and their quality is defined in a way that is, by and large, independent of other formal and functional aspects.

So, then, we now find ourselves in a situation in which design is related to the proximity of the object; proximity, that is, in the strictest sense of the term (with regard to tactility) but also in a wider meaning – as the approach to a process of elaboration of sensorial data in which the colour is 'closer' to the object than the shape because awareness of it calls for a lower level of mental process. The recent appearance of a whole range of coverings and finishes that can be loosely categorized as 'high touch' surfaces shows how the two cultures of design and market demand are both beginning to show a new awareness of the sensorial quality of things. But this is a sector in which there is still a very long way to go, starting off with the definition of a vocabulary of terms with which to talk about it and going on to the creation of cultural instruments upon which to base design choices.

Surface and Time

How long an object lasts has to do with the characteristics of its component parts. But it is the surface that shows all the signs of wear and tear, and it is the surface, too, that lets many of the factors through that contribute to the deterioration of the interior. With the passage of time, interplay of the chemical/ physical characteristics of the article and the agents present in the environment may lead to corrosion, oxydisation, scratching, splitting, and biological decay caused by the organisms present in the material – to say nothing of the normal buffetings of everyday use. The more or less gradual process of decline or disintegration is represented by a downward graph curve generally known as the 'degradation curve'.

The question 'How long will this last?', an important feature of any planning or design procedure, presupposes another: 'How much deterioration is acceptable within that arc of time?', the answer to which is to be sought on two different planes, the cultural and the technical. On the latter the answer is relatively simple and straightforward: the initial quality of the object and of whatever protective coating its surface may have been given must never fall below the level established at the design stage. But the cultural aspect is rather more complicated. Whereas the deterioration of certain materials is culturally acceptable, (traditional building materials, like brick, become 'mellow' with age), for others – for the most part those of more recent development – it is not. A plastic chair, for instance, is either in perfect condition or only fit to be thrown on the scrap-heap. The quality of the surfaces, on the score of protection, takes on, in such cases, a cultural dimension too, (although relatively unexplored, in relation to the ageing process; how to last a long time without showing loss of quality.

Certainly, very few of the newer materials have the virtue of this ability to age with dignity. The new equation 'new material=ageless product' is not necessarily and always true, and the shrewder designers have begun to check up on the 'new materials' expressive capabilities over a more prolonged period of time. But only a start has been made, and first of all a cultural approach is indicated rather than working on the materials themselves. The aim should be to move onward from the recent tradition of design, which has generally seen objects as being frozen in the image of that which is new, to meet up with one of the more resistant bastions of the Modern; the exorcism of death and decadence in a dream of eternal youth.

Amid the profusion of objects that are churned out by present-day production lines only to hurtle at a dizzying pace straight for the scrap-heap, we should be studying the prospect of adding others that have the ability to age and at the same time not to decay, which can stimulate the collective memory and act as functional and analogical slowers-down of time, the passage of the years being mirrored in their mutations.

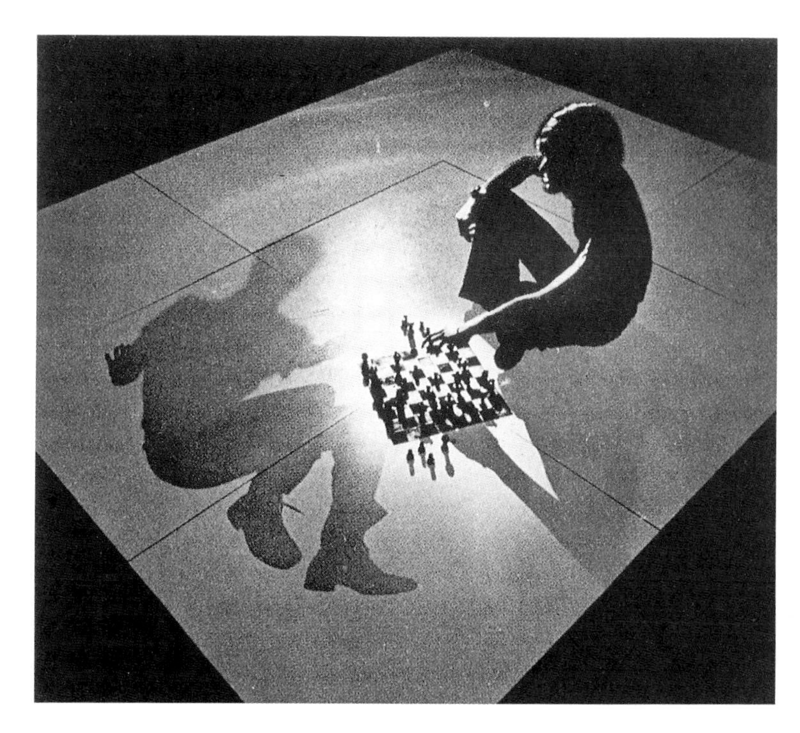

The shadow of a presence on a 'Lumiphos' photoluminescent plastic laminate designed by Clino Castelli for Abet Laminati. Italy 1987.

Reactive and Expressive Surfaces

The time during which surfaces come into contact is not be measured only in a linear dimension, that is to say the time that deterioration takes to leave its signs of entropy. Time is also the variable by which rapid and reversible variations of condition are measured – the time of inter-activity between surface and environment. In fact, one of the most heated arguments in the sector nowadays concerns surfaces' propensity for carrying traces of movement across or pressure upon them (reactive surfaces) or for revealing changes that have gone on beneath in the body that they cover (expressive surfaces).

It is rather as if the tendential loss of relevance of the third spatial dimension were being made up for by the introduction of a fourth temporal dimension. A phosphorescent surface that retains in the darkness a light that once shone and is now no more – a surface with a liquid crystal pigment that changes colour according to variations in temperature – a surface that retains the imprint of whatever has rested upon it – these are just three examples of a range of super-surfaces that have been made possible by the introduction of dynamic qualities at a very high aesthetic and emotional level.

But the future of the reactive and expressive surface is something else again. In the examples that we have just quoted simple types of covering are used to solve equally simple probems. But the bi-dimensional component entering as a feature of other more complex systems can grow into something far, far more important and complicated, offering countless combinations of logical function and emotional and aesthetic consideration.

The most familiar and widely-used reactive surface, sending signals from the user to the internal mechanism, is that of the membrane keyboard, to be found nowadays not only in office machines but also in household electrical appliances and even in toys. Its novelty lies not so much in the membrane keyboard itself nor in its role as interface so much as in the changed appearance of the object that it forms part of. A traditional-type keyboard is a solid and conspicuous physical presence. The membrane keyboard is only a surface layer, the sensorial instrument of tactility diffused inside the appliance, the first step towards a tactile sensitivity which is to be seen no longer just in the man/machine direction but also in the reverse flow. On the other hand, artificial sensoriality offers possibilities that are far more sophisticated. Pressure sensors have been around for quite some time, but have only recently been developed in membrane form. The need to make robots' 'hands' less clumsy urges the use of thinner and lighter materials with ever greater possibilities of approximating the sensitivity of the human touch. Piezoelectric film and conductive rubber (used individually or a mixture of both), are leading to the manufacture of a kind of artificial skin which will be used, however, not only for robots' hands. Imagine a tennis court that can 'feel' when a ball is out – and register the fact, too. Imagine a plate that can trace the outline of an object has has been placed upon and then removed. Imagine orthopaedic examination devices that can reproduce in diagram form the pressure of the human foot.

So the day is not far off when we will be able to remove those inverted commas before

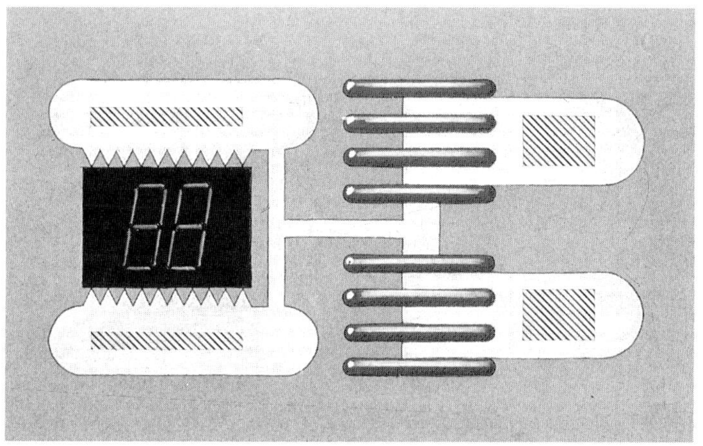

Perry King and Santiago Miranda. A constituent element in the design of a control keyboard of a telecopier for Olivetti. Italy 1980s.

and after the term 'sensitive skins'. Skins have already been made that can 'feel' and recognize surface textures, that can 'feel' and measure the specific heat of a body put in contact with them. 'Skin' has even been produced to resemble human skin in its structure, that is to say with dermis and epidermis, each having its own specific properties and functions.

The evolution of the surface's expressive capacity is progressing on similar lines. Lamps, LEDs, displays and cathode tubes have been showing us for a good while now what is going on inside the devices that they form part of. But these indicative instruments are tending towards a two-dimensional conformity, too. A light bulb can be replaced by an electro-luminescent film that is less than a millimetre thick. Displays, too, are getting thinner and thinner, and much more flexible. And cathode tubes are on the way out, being replaced by flat screens. Where are all these devices going to end up if not on the surface of other objects?

In 1882, the Rev. Edwin Abbott wrote *Flatland*, describing in great detail life in a two-dimensional world. But little did he dream that what he was describing would, in a hundred years time, be one of the major trends of technological evolution. But Flatland was just a flat world, whereas the new two-dimensional world of our own time forms a layer over the old three-dimensional one – a skin.

The Sensitive and Communicative Object

Living organisms without a structure do exist in nature, but there are none without skin. In fact, there are some that are nothing but skin; at the very lowest level of the biological scale there are micro-organisms consisting simply of a membrane dividing an interior from exterior. Skin, with its particular characteristics, is the site of the exchange of

Perry King and Santiago Miranda. The diagnostic zone of the keyboard of Olivetti's Copia 900 photocopier. Italy 1980s.

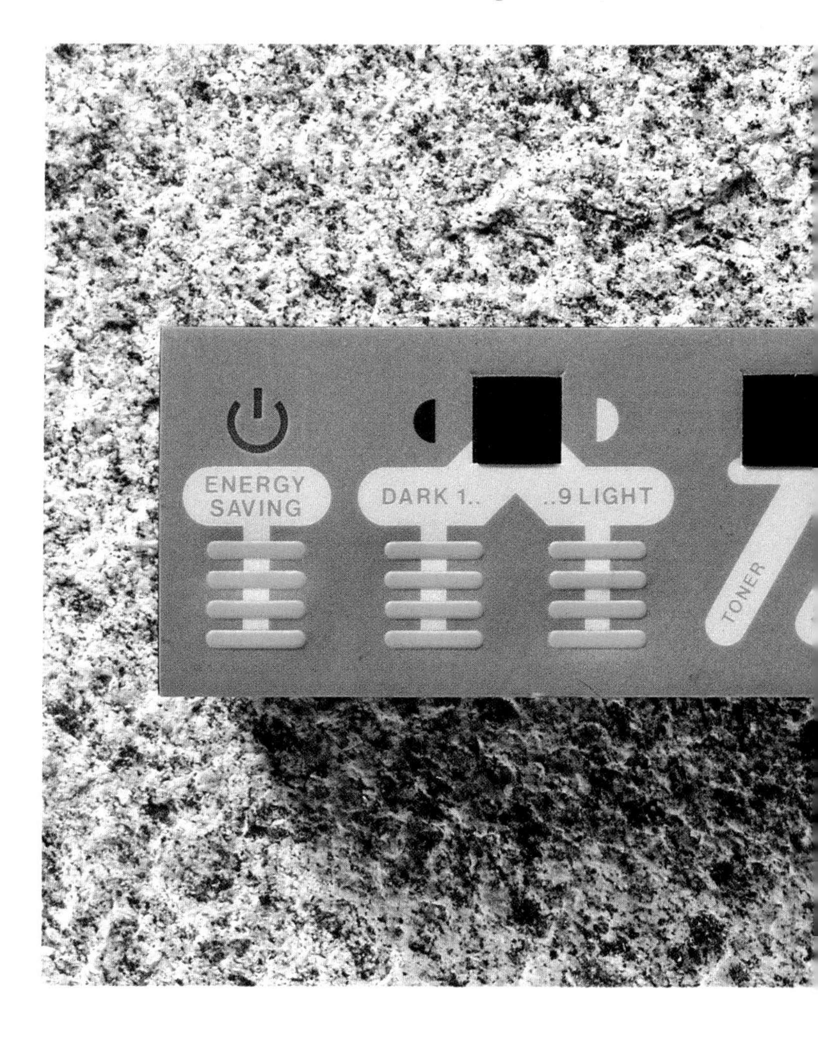

energy and communication that characterizes life.

It follows, then, that when technological advances lead to the production of something artificial very closely resembling the organic world in all its complexity, the surface of that object takes on greater importance and significance, becoming an interface, a filter, the place where those exchanges of energy and communication occur. Technology reaches this point via a route that follows the same pattern as that of biological evolution. The latter starts off from the membrane and ends up with rigid structure organisms, (the vertebrates). The former starts off from simple structures and progresses to an appraisal and exploitation of dynamic surfaces. But from this juncture on the subject branches out to cover so much ground that it cannot be taken further in this article. The evolution of reactive and expressive surfaces leads to a whole new generation of sensitive and communicative objects for which the central point of the design is no longer the physical shape but rather the form of the relationship set up with the environment. These are objects that are defined by their 'behaviour' and their 'personality', and, for this reason, exist on quite a different level from anything that has been produced by man hitherto.

And this, of course, is another question.

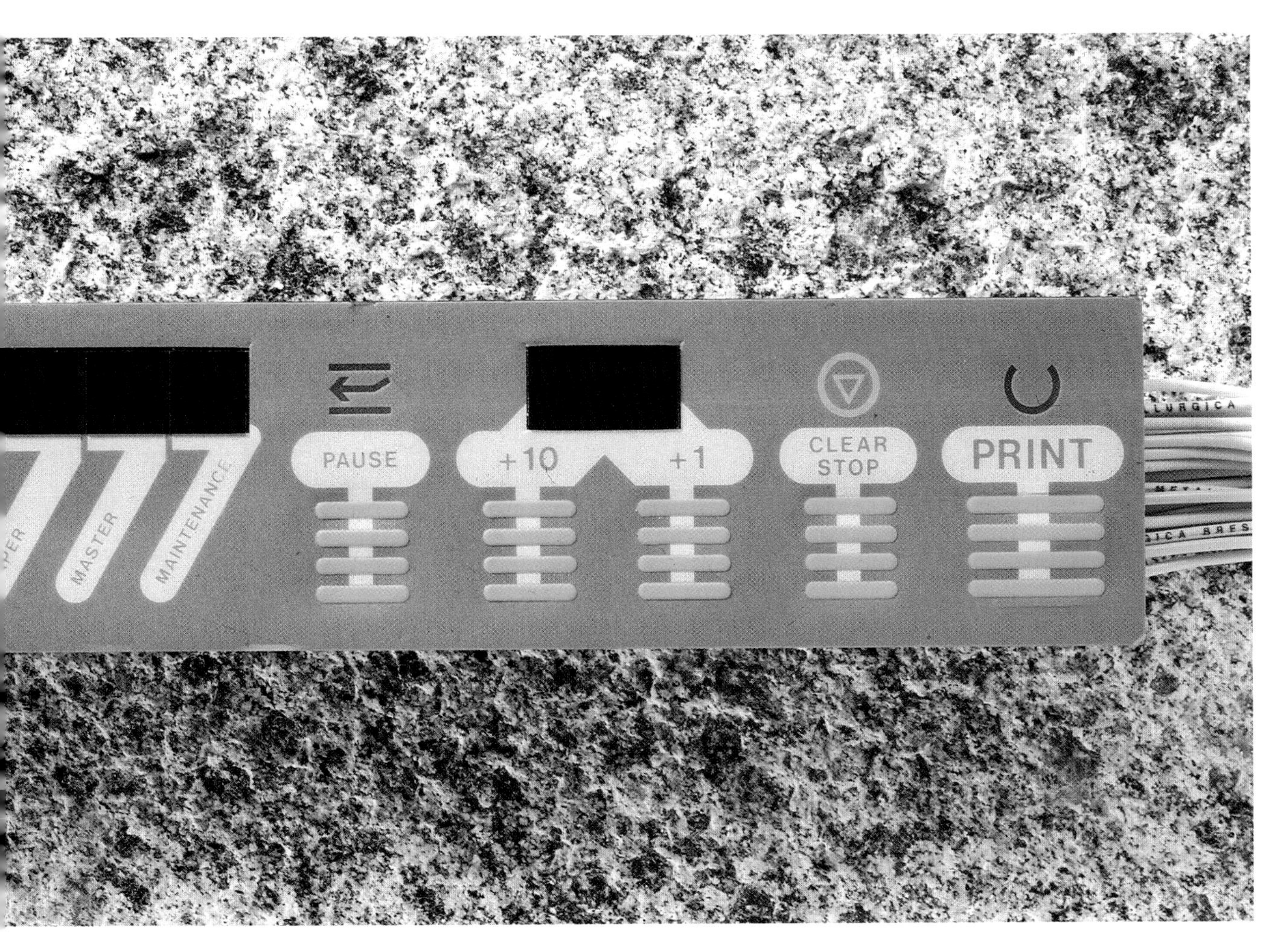

127

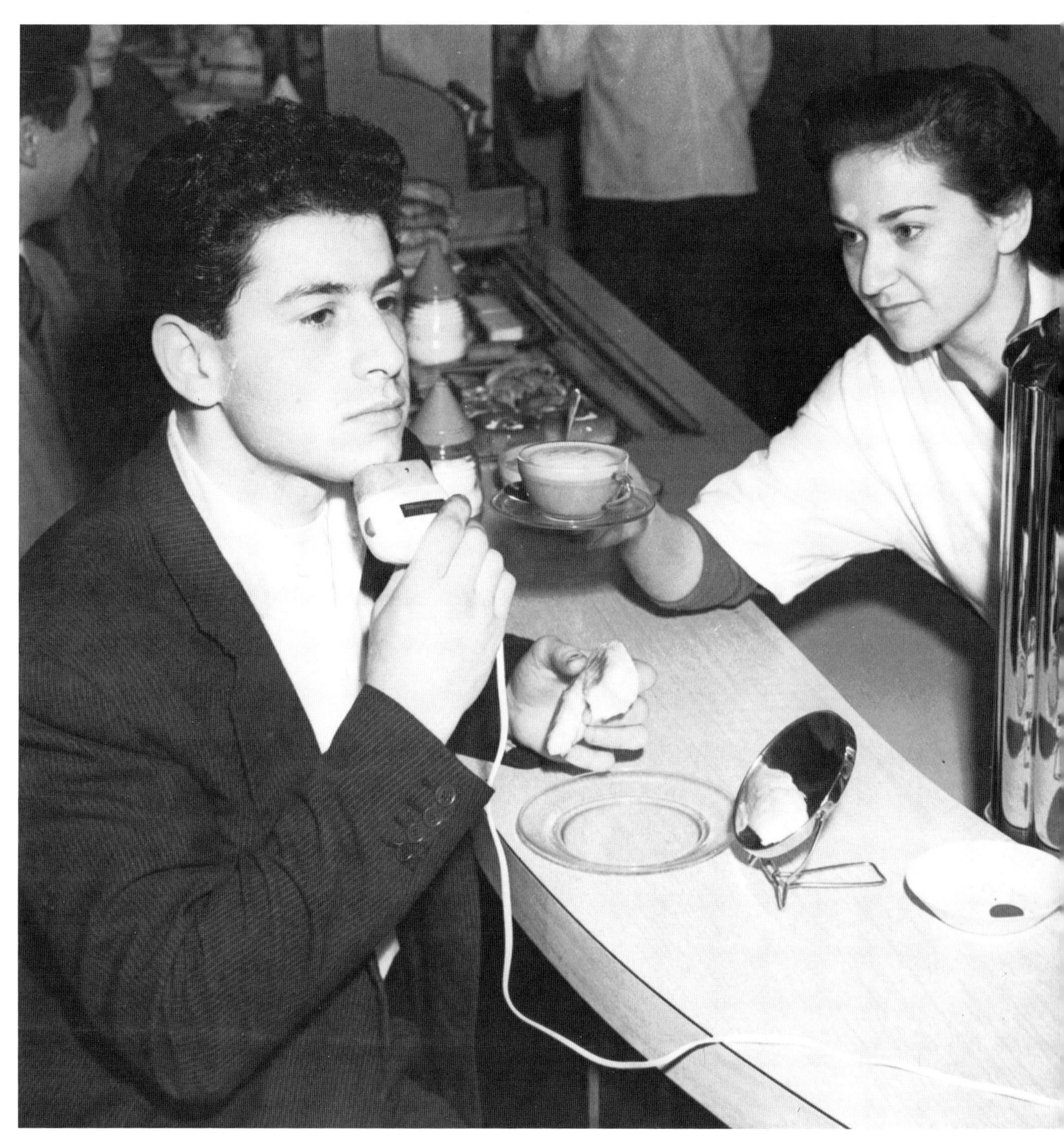

Coffee-bar in Frith
Street, Soho. 1954.

PLASTIC LAMINATE

Barbara Radice

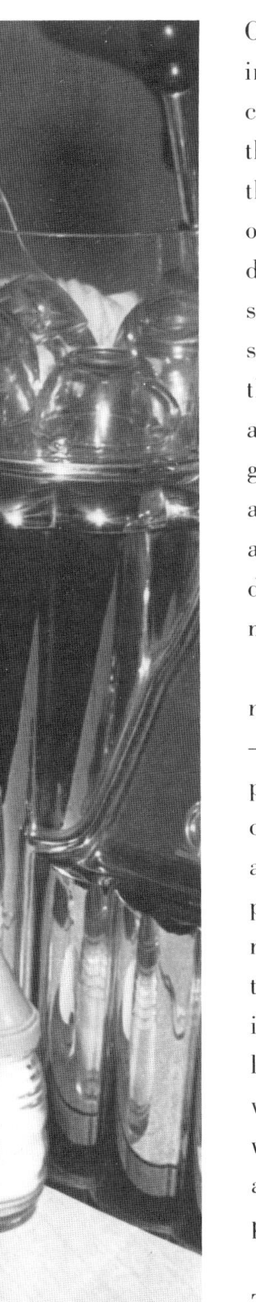

One of Memphis's most important innovations, perhaps their greatest contribution to the turn of style that changed the face of contemporary furniture, has been the use of plastic laminates, and particularly of decorated plastic laminates, in furniture design. This seemed to some people a secondary issue, a matter of intellectual snobbery, a cheap trick, without thinking that the problem of materials in design is anything but marginal or incidental. The greatest changes in taste or style in architecture and in the applied arts have always been accompanied, or better yet defined, by interest in new available materials and technologies.

Using different materials provides not only new structural possibilities, but – above all – new semantic and metaphoric possibilities, other modes of communication, another language, and even a change of direction, broadening of perspective, appropriation and digestion of new values and the concomitant rejection of traditional structures that renewal always involves. With Memphis and plastic laminates, the renewal was generated by a violent switch of cultural context combined with, and magnified by, the introduction of an absolute novelty: surfaces decorated with patterns of the designer's own invention.

Plastic laminates certainly are not new. They appeared years ago on the tables and chairs of bars and coffee-shops, and today rule supreme in ice cream parlors, milk bars, movie theaters, take-out restaurants,

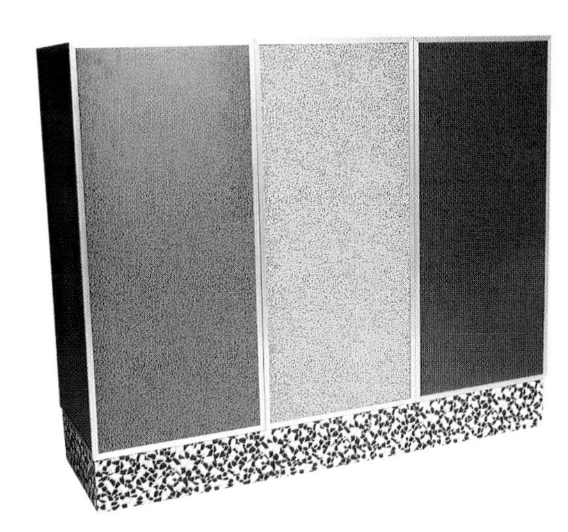

Right: Ettore Sottsass. Wardrobes covered with plastic laminate. Studio Alchymia. Italy 1980.

MacDonalds and Wimpeys. With their sugary colors and fake wood, brick or wicker squiggles, they have become part of the mass urban scene, a symbol of suburbia, that anonymous hinterland, a little naive, a little desperate, but optimistic, positive and self-assured, which to define itself shrugs off both the culture of the city and that of the country, making its myths of modern materials, naturally absorbing all technological shocks, identifying body and soul with the future, knowing it 'is' the future.

Plastic laminates made their way into the home years ago on account of their 'practical and functional' qualities. Today they can even be found in the homes of the wealthy, but most of the time they are 'hidden away' in closets, bathrooms, and kitchens, or at best in the children's room. They have never appeared in entrance halls or living rooms, the 'formal' rooms entrusted with displaying the owners' status symbols and prestige. Plastic laminates today are still a metaphor for vulgarity, poverty, and bad taste. As a

Detail of 'Antibes' showcase designed by George Sowden. Memphis. Italy 1981.

result they are also excluded from those public places that aspire to a certain standard of 'elegance' be they restaurants or bars, night clubs, confectioner's shops or boutiques.

Memphis turned this situation upside down. It took plastic laminates and put them into the living room; it studied and explored their potential; it decorated them and glued them on tables, consoles, chairs, sofas, and couches, playing on their harsh, noncultural qualities, their acid-black corners, their ultimately artificial look, and the dull uniformity of their surface, which is void of texture, void of depth, void of warmth. And yet, as Emilio Ambasz has pointed out, these laminates are 'forever young, eternally vibrant'. The greatest novelty of Memphis's plastic laminates is their decoration and the most important feature of this decoration is

Punks. England 1983.

its anonymity, its absence of signs, of quotations or metaphors associated with codified culture. The iconographic package of Memphis decorations comes, like the laminates, from unorganized cultural areas such as suburbs or growing cultures. The patterns are graphic formulations of brutally decorative geometric motifs, which in some instances, like Michele De Lucchi's 'Micidial' and 'Fantastic' patterns, even have names that recall the emphatic and paradoxical atmosphere of certain comic strips. Others evoke stereotypes of false Venetian blinds, false meshes, false serpents, even false masterpieces of painting. Or, as in Ettore Sottsass's now familiar 'Bacterio' and 'Spugnato' patterns, the laminates evoke neutral and anesthetizing organic forms. Patterned plastic laminates have been so important in the definition of New Design that their birth coincides with the research into new types of furniture undertaken by Sottsass following the 1977 agreement with Croff Casa. The first drawings for the 'Bacterio' and 'Spugnato' laminates date from this period and were later produced by Abet Print and used for the first time on the experimental furniture made for Abet Print and shown at Studio Alchymia in 1979.

It seems that the idea of patterned plastic laminate furniture came to Sottass as he was drinking coffee at ten o'clock one morning at the pink-and-blue veined counter of a quasi-suburban milk bar near his house – a place frequented at that time of the morning by post-office employees and old ladies looking for cats to feed. Equally fortuitous and slightly decadent in a Klimt-like way, was the inspiration that struck Michele De Lucchi for his first pieces in Memphis laminate: he was watching teen-aged punks with talc-white faces painted in bright colours on New Year's Eve 1980-81 in Trafalgar Square.

Both were instances of lightning inspiration, explosions of awareness only possible in brain cells already prepared for a certain chemical reaction, already hyped for a solution that had been a long time coming. And both were cases of socio-linguistic inspiration, of messages more anthropological than decorative. This is probably why Memphis decorations had such a forceful and immediate impact in so many different fields. Soon after their début in furniture design, they appeared on T-shirts and sweaters, as a graphic support in magazines, and even printed on a famous new-wave brand of shoes. Given the demand, Abet Print has put three designs in its general catalogue and is preparing a separate Memphis catalogue with patterns by Sottsass, De Lucchi, and other designers chosen by Memphis in a workshop on decoration set up by Sottsass at the Kunstgewerbeschule in Vienna.

Ettore Sottsass. 'Bakterio' plastic laminate. Abet Laminati, 1980.

AND OF PLASTICS?

Ezio Manzini

Detail of the keyboard of the 'Valentine' typewriter designed by Ettore Sottsass for Olivetti. Italy 1969.

Our eye travels over the objects in our daily surroundings. Forms with a certain quality, a quality deriving from the materials they are made of. Memory, experience and intuition produce their names from a mental catalogue: wood, iron, plastic . . . Our relationship with the material physicality of things passes through this ability to recognise what objects are made of.

Now, however, there is an interference in this process of recognition; memory, experience and intuition cannot help us any longer: with products of the most recent generation we can say more what they seem to be made than what they are really made of.

This is not merely the result of unfamiliarity with the new. It has deeper roots: in an age in which science and technology work at the micro-level products no longer seem to be made of matter capable of providing a series of material data to our senses. Instead, we have a continuum of possibilities: performance and look may now be found in the most unexpected and unclassifiable combinations.

Vico Magistretti. 'Gaudi' armchair manufactured by Artemide. Italy 1970.

Hence the inevitable tendency to distinguish between what a material is (on the chemical/physical level) from what it seems to be. This relative autonomy of image from 'material reality' is not particularly surprising. It is only part of a larger phenomenon of technological, social and cultural evolution in which the 'apparent' becomes the only reality to which one can refer.

Recent developments have therefore critically undermined one more basic tenet of modernist thought: the idea that 'materials possess their own genuine image'. This is not (or not exclusively) the result of a cultural choice that deliberately values ambiguous 'fake' above 'sincerity' in materials. No, the critical undermining of the idea of a material's genuine image comes more from the very properties of the new materials, available in an extremely wide range of forms and hence able to possess 'sincerely' a multiplicity of images.

This new situation has been generated by the technological, economic and cultural process, accelerated above all by the appearance and development of plastics. The versatility of these materials has stimulated competition between them, resulting in a rapid multiplication and extension of properties.

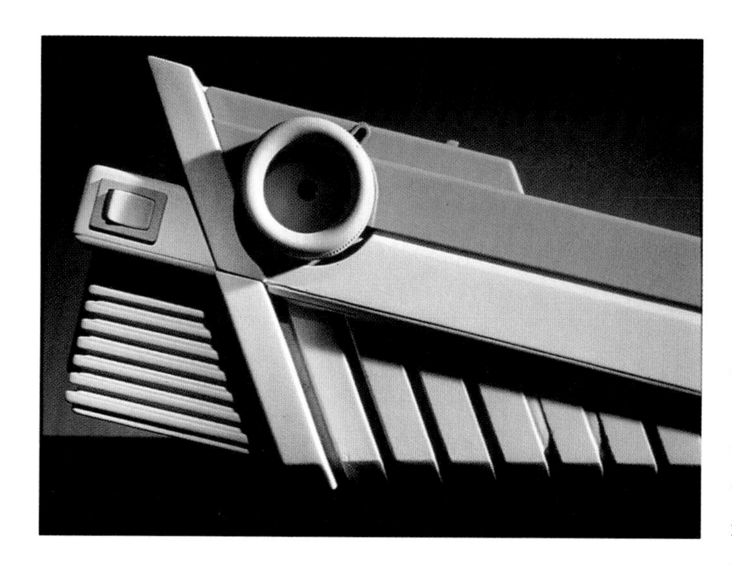

Mario Bellini. ET 55P typewriter manufactured by Olivetti. Italy 1987.

'Valentine' typewriter designed by Ettore Sottsass for Olivetti. Italy 1969.

Their artificiality and lack of history have helped to damage, and threatened to demolish entirely, a whole system of images and values based on natural qualities and consolidated by a long perceptual and symbolic tradition. The history of plastics is the history of this 'destabilisation'. It is not a simple and straightforward story: the word 'plastic' has itself undergone gradual changes in connotation until the point reached today in which it has no clear affective meaning. Various assorted images have alternated without ever eliminating one another: the exotic; the everyday; vector of progress; threat to the environment; the aerospace component; the supermarket bag; the kitsch; and the highly sophisticated design. The word 'plastics' now evokes contradictory connotations, creating an ambiguity that weakens the word's communicative value.

An equally complex story is involved in the relationship between plastics and the world of design. Plastics have exploited their formable qualities not so much in the direction of fulfilling technical and constructional needs as in the expression of different images. At first the simplest solution was to imitate 'nobler' materials.

Subsequent criticism from the Modern Movement was inevitable but basically contradictory: plastics were placed on the same level as traditional materials, but it was not realised that flexibility of image is one of their basic characteristics: hence, paradoxically, for plastics even imitation is a form of 'sincere' expression. But the great cultural importance of the Modern Movement meant that plastics had to accept its conditions: in order to enter 'high culture' (hence also large-scale manufacturing) they had to give themselves an independent image, their very own 'sincere image'.

After the war, then, plastics looked for and found such an image. It was not easy to get it accepted: it was a card played bravely and decisively by only a few designers and very few manufacturers. Success was hard-won and well-earned by these pioneers.

Plastics won a place in the catalogue of accepted materials when their economic qualities combined with their cultural qualities. Distinguishing marks of the new

Marco Zanuso and Richard Sapper. Algol 11 television set manufactured by Brionvega. Italy 1964.

Overleaf: 'Dalila' chairs in rigid polyurethane foam, designed by Gaetano Pesce for Cassina. Italy 1980.

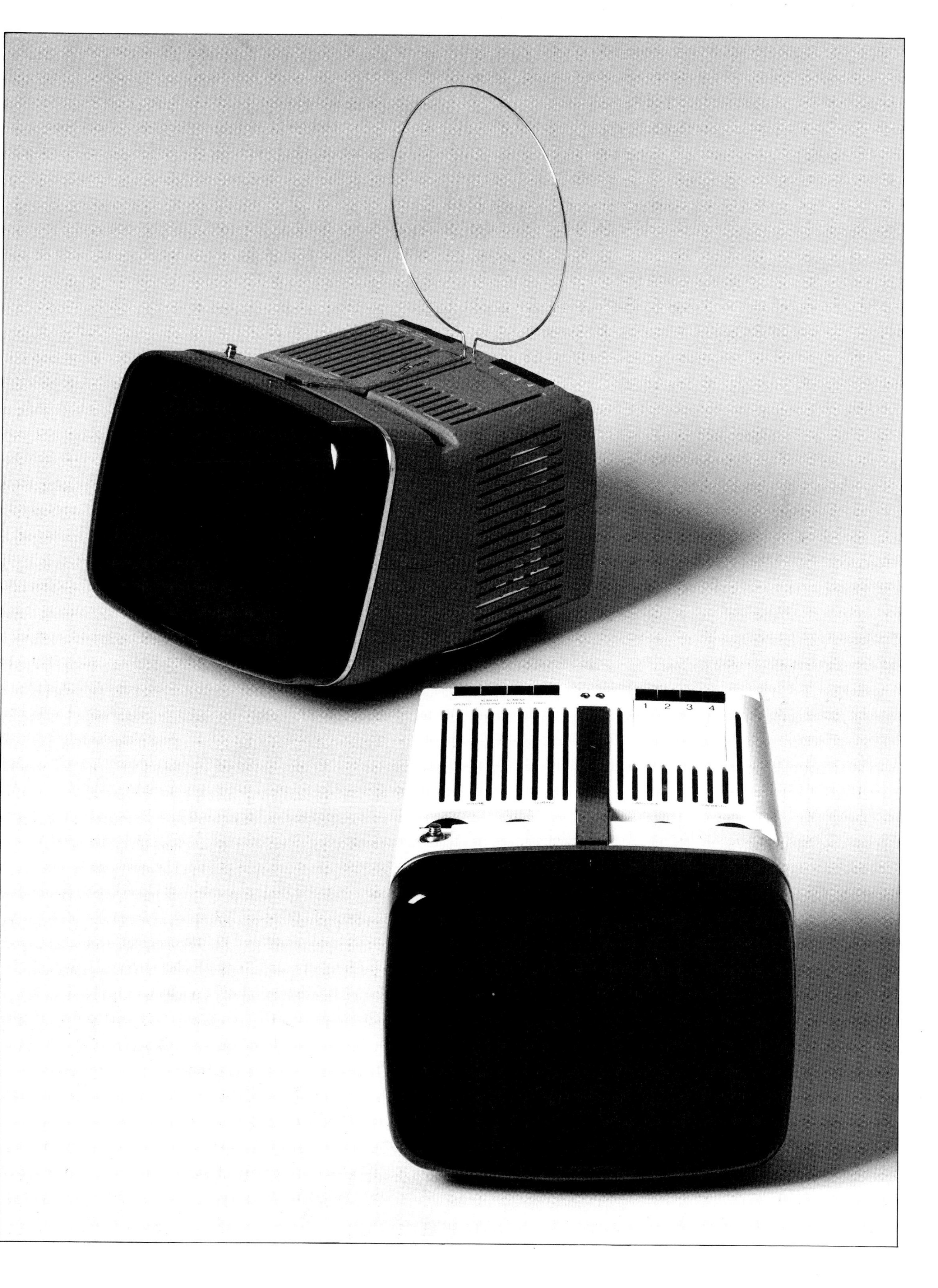
1 2 3 4

'Nesso' light designed by the Gruppo Architetti Urbanisit Citta Nuova for Artemide. Italy 1964.

image: clear and moulded forms, rounded junctions, primary colours, generally shiny surfaces and only one material for the whole product. With this identikit plastics take their place in the collective memory alongside existing materials. Then this favourable atmosphere changed: the energy crisis and concern about the environment (though frequently inaccurately expressed) put plastics in a bad light in the second half of the 1970s. Evolution in taste subsequently robbed plastics of the up-to-date shine on their hard-won image. People now put aside the ingenious fantasies about worlds of plastics created from the infinite repetition of a system of simplified forms. The debate (and we are now near to the present day) shifted from interest in materials to an interest in aspects of language and communication.

An analysis of the most recent design developments shows that themes like 'complexity', 'fragmentation', 'quotation' and 'hybridation' can only be translated into products because the material used possesses a 'formability' and adaptability without precedent in the history of design. We see that stressing of the communicative aspect of products has been made possible by the availability of materials that adapt themselves to the syntax of the design in the same way that the words of a language adapt themselves to the syntax of a text.

Plastics filter into this new universe of images through a thousand pores, producing every kind of form, carrying an extraordinarily wide variety of meanings.

We are unable to present a precise image for plastics in the 1980s and 1990s – not because they are so little used, but precisely

'Pausania' light designed by Ettore Sottsass for Artemide. Italy 1983.

because they are used so widely as to make formation of a clear impression difficult.

Plastics have grown enormously in number, have increased in importance beyond all expectations, they have been combined in use freely with a wide range of other materials, and have been largely freed from ideological and cultural limitations. They can now fully develop their specific qualities of adaptation and imitation and continue their gradual penetration of the system of products. Paradoxically (at first sight), with this excess of qualities, with this extraordinary and kaleidoscopic representation in the world of products, plastics have lost their former specific identity. A situation not really different from that described by Musil when he writes 'The man without qualities is made of qualities without the man.'

Overleaf: Anna Castelli Ferrieri. Range of ABS tables, model 4310, manufactured by Kartell, Italy 1983.

'Seconda' chair
designed by Mario
Botta for Alias.
Italy 1982.

'Light-light' chair
designed by Alberto
Meda for Alias.
Italy 1987.

'Tizio' light designed
by Richard Sapper for
Artemide. Italy 1972.

PLASTICS IN THE '80s

Sylvia Katz

sten Jorgensen.
ad bowl, model
1, and servers,
del 2522, designed
Bodum.
tzerland 1980s.

During the last five years there has been more progress in materials development than in the previous twenty. Such is the scale of the plastics revolution. The image of plastics is constantly re-focusing. The bright and colourful Sixties are gone and the new plastics are 'high-tech', sophisticated, innovative materials. Our lives are governed by them. Although created by other technologies, the whole sphere of telecommunications, information processing and medicine could not function nearly as efficiently without plastics.

Plastics thrive in a synergetic atmosphere, and once the connections have been made, totally new products emerge. Following the commercialization of the silicon chip in 1959 plastics have been the chief medium for giving form to computers and microprocessors, sensitive membrane keyboards, liquid crystal displays and flexible printed circuits. Circuits laminated into plastics film have created ultra-slim calculators and electronic products; 'intelligent' memory chips embedded into flexible plastic film cards has resulted in portable access to a tremendous amount of information.

The medical Care-Card was the first 'smart card' to be launched in Britain in 1989 following experiments in Europe. Its 16-kilobyte memory chip may not look much but it can hold a file-full of life-saving information. It is another step in the direction of the paperless practice and possibly more important in the end, part of the process of the patients as people, not victims of a system. For the first time since medical records began they will have access to their own files, and will be able to go into the surgery and scroll through their own medical history.

Plastics have broken down the traditional categories of materials by offering new textures, colours and moulding possibilities which in turn have altered our perception of the products of the late twentieth century. The exciting sculptural clothes by the Japanese designer Issey Miyake are constructed from industrial synthetics not normally found in high fashion although they have been available for many years: helmets of silver polyurethane foam sheet and bustiers of moulded silicone. Plastics have changed the designer's vision. 'The old rules have gone', declared the leading French designer Philippe Starck in 1987, 'And for the moment we are in a situation where innovative ideas from anywhere have an equal chance of being heard'.

Natural and synthetic rubber are other examples of a material once confined to industrial applications but which have been given new life since their discovery by designers. Now rubber can be found moulded into solid desk accessories by the Italians; colourful shoes and jewellery by the Japanese; or non-slip inserts into tableware by the Swiss.

Ironically the chief shortcoming of plastics has been their own versatility. What other substance can assume so many guises?

Moulded car dashboard using a variety of plastics. Bayer 1980s.

They vary in form, colour, texture and smell; they can appear soft, hard, flexible, opaque, transparent or translucent, or in the form of a foam, film, gel, blend or an alloy; they can be conductive, polychromatic, fluorescent, or autonomous with ready-to-fix inserts. It is because of this versatility that is has taken so long for this chimeric family of 'magical substances' to be accepted, let along understood. It is not easy to relate a Coca-Cola bottle to a polyester shirt, yet they belong to the same family of polymers. In addition, designers have been surprisingly slow to exploit these chemical materials.

The major persuasive factor of plastics is that of substitution, but plastics go beyond normal expectations and outperform traditional materials. A bottle moulded in glass is transparent, cheap, recyclable, hygienic but shatters dangerously. A bottle moulded in PET (polyethylene terephthalate, a polymer that prevents the bubbles in a fizzy drink from escaping) is all of these and more because it bounces, and even the massive three-litre-size can be handled safely by a child. PET is also soft and warm to the touch, and it sparkles in a way not possible with industrially moulded glass – a property that makes the liquid inside especially appealing.

Even more sophisticated is a bottle made of several layers of different types of plastics, each with its own specific function. A bottle blow-moulded from a sandwich of polypropylene ethylenevinyl alcohol (PP/EVOH/PP) has many of the properties of the PET bottle listed above but it is squeezable as well. Five layers of this multi-layer

The GE Vector 1 car. GE Plastics. USA 1988.

extrusion – a kind of polymer lasagne – (PP/EVOH/PP/EVOH/PP) – produces a high-barrier plastic which is now used to make food trays which can be 'shelf stable' at room temperature, and later put in a microwave oven. No energy is wasted keeping the food refrigerated. Applications like these are helping to encourage a belief in plastics as quality materials of the future. The PET bottle is probably the best-known example as it went beyond simply creating a new and immense market for plastics bottles, but promoted plastics as a viable alternative to glass.

PET was first developed by ICI in 1941 as a fibre and later as a sheet. A blow-moulded PET aerosol can, known as the Petasol container, was launched in 1989 by Fibrenyle and this will not only replace the traditional metal can but allow designers freedom to change the shape of future aerosols.

Blends and alloys in plastics are a phenomenon particular to the 1980s, and it seems that almost every day new materials and uses are developed. In the medical world nylon is a favourite bio-material but it can become brittle. Monsanto developed a nylon/ABS alloy tough enough to make a 24-hour cardiac monitor weighing only 250g, including cassette tape and batteries.

We have made plastics that can pass from −40°C in the freezer to 230°C in the oven, look good on the table, withstand the aggressive treatment of a dishwasher, and then go through the cycle all over again – in other words, a ceramics susbstitute with both lightness and toughness. Not long ago radar-invisible planes and aircraft wings constructed from plastics seemed improbable, but that was without the knowledge of today's composites and super-polymers.

Philippe Starck. Richard 111 armchair manufactured by Baleri. Italy 1980s.

Electronic cordless personal scale with remote LCD display, model 9255. EKS. Sweden 1980s.

Plastics have a habit of springing surprises on us. Just when we begin to think that we understand polymers, the shape of the future changes again. After having been for years the best electrical insulator, plastics now appear to be offering us the opposite property – conductivity. Research began in the 1980s on designing conductive components to order, and products which at present are designed around their electrical circuits will evolve new forms, and undoubtedly new applications.

Video Walkman. Sony. Japan 1989.

We have also invented plastics that can withstand temperatures from −260°C up to +260°C. And the plastics (polyacetylene) transistor has arrived too. These are miracle materials indeed, and our children will soon be saying how quaint it is to see a kettle, iron, or even in the near future, an entire car, made of metal.

The late 1980s have seen a dramatic change in life styles, a move into the era of telecommunication and information technology, described by John Thackera as the 'global network of invisible forms that has come so quickly to dominate our lives' (MODO 112-9). Plastics are the supremely flexible media for communicating intelligence and ideas. When your workplace is wherever *you* are, at home, in the car, or in a taxi, portability is paramount, and the market for miniaturisation has encouraged a mass of gadgets for the nomadic telecommuter: high-tech datacards, calculators and lap-top processors, credit-card sized mini-kits holding stationery, tools, sewing equipment, first-aid and, of course, executive golf accessories. There is even a shop in Tokyo specialising in miniaturised business equipment designed to fit the briefcase of the travelling 'salaryman'. From the first Walkman of 1979 to the 1989 paperback-sized Video Walkman, plastics have easily

The Junior Computer. Team Concepts Electronics Ltd. Hong Kong 1980s.

Interlocking salad servers. Smart Design Inc. New York 1980s.

adapted themselves like a chameleon to innovation in every area of our changing life styles.

Of course these developments are not all geared to the metropolitan man. Spin-offs improve the quality of life for everyone. In the wave of the new-generation of electronic toys that have been flooding in from the Far East are valuable educational ideas, all packaged in plastics. The system developed by Video Technology Electronics in Hong Kong, called 'Socrates', uses colour animation to present over ninety programmes for children aged five to ten years, and covers maths, spelling, word problems, music and drawing skills.

Before the century runs out a new lexicon will have to be created for plastics, a terminology that can be understood in a direct, non-technical way. New words are already needed to describe the qualities and use of plastics things: 'patina' and 'inlay' are

words adopted from the language of other technologies, even the term 'graining' can be used to describe a decorative device, but the sheer *plasticity* of late twentieth-century artefacts is something essentially synthetic.

Despite the growing sophistication of plastics designs, knowledge about the materials themselves and the education given to designers and engineers is badly out of proportion to the amount of plastics that surround us. Everyone knows that wood grows on trees but how many can say where plastics come from? Or describe the basic processes? By the end of the 1980s ideally each of us should have an understanding of the simplicity and beauty of the concept of countless identical groups of molecules that link up to form the enormous chains known as 'polymers' which in turn, modified by various additives according to a particular recipe, are moulded into 'plastics'. How sad that such an apparently mundane word should describe a magic organic process. The metamorphosis of primeval sludge from beneath the earth into shining multi-coloured objects symbolises Man's power to control the atoms and shape the molecules from which both he and his planet is made.

However, the present ignorance about plastics is both to their advantage and to their disadvantage, the advantage being that plastics can continue, as they always have done, to quietly infiltrate and multiply until we and they are inseparable.

But ignorance is not bliss. The problem of not knowing the long-term environmental effects of high-tech polymers has already influenced the market. Witness the 100-lire tax levied by the Italian government on plastics carrier bags, or the deposit demanded in Germany on one-trip plastics bottles. Reverse Vending Machines have already been made that will politely say 'thankyou' whenever a plastics bottle is returned, and a major British food chain has started to educate its customers about 'environment-friendly' packaging. Realising that there can be no successful recycling without a basic knowledge of the different plastics, the company will be selling recyclable PET bottles with special labels, and it has already started the sort of improvements that come easy to plastics, such as braille labelling and tamper-resistant tops.

Wood grows in its own natural way. Ceramics and metals periodically make technological leaps. The intriguing fact about plastics is that they are in a constant state of evolution, replacing, and improving on, their own species. Did someone say that we had entered the 'Plastics Age'? Why, we've only just started the journey.

Lipstick vacuum bottle, model 1600. Bodum. Switzerland 1980s.

CONCLUSION

As the compromise between what is 'possible' and what is 'thinkable' moves through this century it carries with it the load of its trajectory. Thus the 'meanings' of plastic products are, at any one moment, dependent upon both their contemporary cultural context and their past. This complexity increases our essentially ambivalent relationship with them: Thus while we accept that without plastics many of the achievements of advanced medical practice would be impossible (heart transplants, joint surgery, neo-natal care etc.) and that the increased safety and hygiene offered by, for instance, children's toys and baby care items are dependent upon the new materials, we still have an underlying fear that somehow plastics *are* cheap and nasty and in some way inferior to the 'real thing'.

The key to the problem of plastics lies in their ubiquity. So widespread are they in today's environment that it is inevitable that they are both loved and despised. A quick scan of any milieu will show that plastics are *the* late twentieth-century materials.

Specific aesthetic objections to plastic products include their weightlessness and their inability to age well. While many brand-new plastic products have bright shiny presences, once they have become scuffed and dirty they quickly enter the arena of 'environmental pollution'. Equally, while plastic packaging on the supermarket shelf extends the life-span of its contents and is perfectly acceptable, the same packaging discarded in the countryside presents an environmental threat, and earns our unanimous disapproval.

The 'lightness' criticism is based on an underlying assumption that weight, substance, quality and value are necessarily inseparable from each other. The lack of weight of, say, a modern vacuum cleaner, typewriter or camera is frequently associated with cheapness and lack of quality. The perversity of this attitude becomes clear when one looks at the situation from another perspective – housewives are offered lighter household tools and small technological goods are rendered both more compact and more portable. The deep-seated conditioning that persuades us that the new lightness is, *per se*, undesirable needs to be overcome if we are to reap the many advantages that plastic products have to offer us.

The ageing problem is harder to overcome but suffice it to say that designers are at this moment engaged in developing new surfaces for plastic products which will resist that inevitability. The responsibility for re-establishing a rapport between plastic products and their users lies at the feet of designers: They have to create a means of making objects 'acceptable' to the public at large. Where high-technology goods are

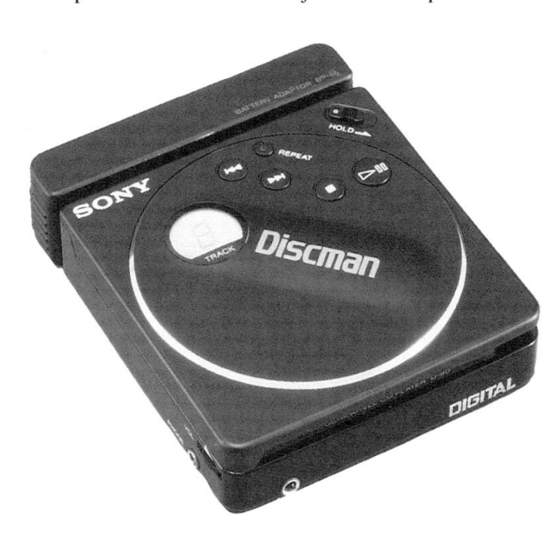

Sony Discman. Japan 1988.

Philips stereo radio-cassette. Holland 1980s.

concerned there is little point in trying to 'explain' their workings to the consumer who needs only to know that they *do* work and that they can improve the quality of everyday life. Equally the technological complexity of plastics themselves mitigates against a 'truth to materials' solution to the problem or an attempt to let consumers understand the difference among diverse materials. Instead the designer can concentrate on the direct sensorial relationship of the object with its user and, increasingly, plastics allow for a more sophisticated treatment of the object's surface, in terms of both decoration and texture. The very flexibility and range of 'possibilities' inherent in plastics means that they can offer an increasingly wide range of sensorial options, re-establishing a direct rapport between object and user. This relationship replaces the more 'rational', functional one so essential to the Modern Movement.

The need to rethink our responses to plastic products is an urgent one. While an understanding of the social and cultural reasons for our inbuilt prejudices and suspicions of these most modern of materials helps to put things into perspective, we still need to move beyond the ethical constraints imposed by the Modern Movement and accept plastics on their own terms. Only then will what is 'thinkable' begin to take full advantage of what is 'possible'.

Penny Sparke

Philippe Starck. 'Dr. Glob' dining chair manufactured by Kartell SpA. Italy 1988.

BIBLIOGRAPHY

This bibliography lists the main book material available in this area. Further research can be undertaken by following up the bibliographical information presented in them and by consulting the specialist periodical material also available.

Allcott, A. *Plastics Today* London 1960

Arnold, L. K. *Introduction to Plastics* London 1969

Astarita, G. and Nicolais, L. *Polymer Processing and Properties* New York 1984

Beck, R. D. *Plastic Product Design* New York 1980

Brydson, J. A. *Plastics Materials* London 1975

Buttrey, D. N. *Cellulose Plastics* London 1947

Buttrey, D. N. *Plastics in Furniture* London 1976

Cherry, R. *General Plastics* Illinois 1941

Cook, J. G. *The Miracle of Plastics* New York 1964

Cook, J. G. *Your Guide to Plastics* Watford 1964

Dingley, C. S. *The Story of B.I.P.* Birmingham 1962

DiNoto, A. *Art Plastic: Designed for Living* New York 1984

Dubois, J. H. *Plastics History USA* Boston 1972

Dubois, J. H. and John, F.W. *Plastics* New York 1981

Duffin, D. J. *Laminated Plastic* New York 1966

Fielding, T. J. *History of Bakelite Limited* London undated

Fleck, H. R. *The Story of Plastics* London undated

Friedel, R. *Pioneer Plastic: The Making and Selling of Celluloid* Wisconsin 1983

Gabka, J. and Vaubel, E. *Plastic Surgery: Past and Present* New York 1983

Gait, A. J. *Plastics and Synthetic Rubber* Oxford 1970

Gloag, J. *Plastics and Industrial Design* London 1945

Gracco, L. S. *Plastic Thoughts* Milan 1986

Groneman, C. H. *Plastics Made Practical* Milwaukee 1948

Hollander, H. *Plastics for Jewelry* New York 1977

I.C.I. Ltd. *Landmarks of the Plastics Industry* Birmingham 1962

Katz, S. *Classic Plastics* London 1984

Katz, S. *Early Plastics* Aylesbury 1986

Katz, S. *Plastics: Design and Materials* London 1978

Kaufman, M *The First Century of Plastics* London 1963

Kellaway, T. W. and Meadway, N. P. *Introducing Plastics* New York 1944

Leyson, B. W. *Plastics in the World of Tomorrow* London 1946

Lomax, G. A. *Plastics and their Use in Craftwork* Leicester 1936

Lushington, R. *Plastics and You* London 1967

Mansperger, D. E. *Plastics: Problems and Processes* Pennsylvania 1943

Manzini, E. *The Material of Invention: Materials and Design* Milan 1986

Megson, N. J. L. *Plastics* London 1948

Merriam, J. *Pioneering in Plastics: The Story of Xylonite* Suffolk 1976

Moore, G. R. and Kline, D. E. *Properties and Processing of Polymers for Engineers* New Jersey 1984

Morello, A. (intro.) *Plastiche e Design* Milan 1987

Mumford, J. K. *The Story of Bakelite* New York 1924

Newman, T. *Plastics as an Art Form* London 1964

Newman, T. *Plastics as Design Form* New York 1969

Passoni, F. *Art and Plastics* Milan 1975

Powell, P. C. *Plastics for Industrial Designers* London 1973

Quarmby, A. *Plastics and Architecture* New York 1974

Redfarn, C. A. *A Guide to Plastics* London 1958

Rondillion, M. *Bakelite* Paris 1982

Roukes, N. *Sculpture in Plastics* New York 1968

Sasso, J. and Brown Jr, M. H. *Plastics in Practice: A Handbook of Product Applications* New York 1945

Simonds, H. R. and Bigelow, M. H. *The New Plastics* New York 1945

Smith, P. I. *Practical Plastics* London 1947

Steele, G. L. *Exploring the World of Plastics* Illinois 1977

Van Doren, H. *Industrial Design: A Practical Guide* New York 1940

Weissler, S. (ed.) *Plastik Welten* Berlin 1985

Yarsley, V. E. and Couzens, E. G. *Plastics* Harmondsworth 1941

Yercombe, E. R. *Sources of information on the Rubber, Plastics and Allied Industries* Oxford 1968

Exhibition Catalogues

B.I.P. Ltd: Newport, R. *Plastics Antiques* London 1977

Milan: Rodolfo II Gallery. Rabolini, A. *Gli Anni Plastici* Nov.-Dec. 1982

New York: Museum of Contemporary Crafts. *Plastic as Plastic* 1968-9

Prague: Museum of Decorative Arts. Lamarova, M. *Design and Plastics* Oct.-Dec. 1972

Rotterdam: Boymans-Van Beuningen Museum. *Bakeliet: Techniek, Vormgeving, Gebruik* May-July 1981

PICTURE CREDITS

Abet Ltd. 114, 119, 120, 121, 122/3, 124, 131b
Alias Srl., photo Aldo Ballo 142a, photo R. Sellitto 142b
Archigram Archives 96, 108, 109
Architectural Association 97
David Arky 36, 38
Artemide SpA, photo Aldo Ballo 94b, 132, 138, 139, 143
Baleri Italia Srl. 147b
Aldo Ballo 100, 101
Bayer UK Ltd., photo Peter Bloodworth 146
Bethnal Green Museum of Childhood, photographs courtesy of the Trustees of the Victoria & Albert Museum 55b, 56
Bodum UK Ltd. 144, 151
Brionvega SpA., photo Aldo Ballo 135
British Architectural Library/RIBA 68
British Library 8, 40, 46ab, 47a-c, 50, 51, 55a, 57, 59a-c, 62a, 63, 66, 72, 110, 113
The Brooklyn Museum, (accession no. 85.9 purchased with funds donated by the Walter Foundation), photo Schecter Lee 49, (accession no. 88, 67) 54
Cassina SpA. 116, 117, 136/7
Studio De Pas, D'Urbino, Lomazzi 99
Design Museum 58, 90, 94a, 95, 115, 118, 130, 153ab
EKS (Sales) Ltd. 148
G E Plastics 147a
Brian Hagiwara 15, 24b, 27, 28
David Harman Powell 89a
Hille 92
Kartell SpA. 10b, 84, 85ab, 140/1
Collection Sylvia Katz 30a, photo Fritz Curzon 6, photo Henning Christoph 10a, 24a, 31, 34, 60ab, 70, 71ab, 111ab
Photo Ken Kirkwood 52
Collection Susan Kelner Freeman, photo Schecter Lee 44/5
King Miranda Associati 91a, 126, 127
Musée du Peigne, Oyonnax 14a, 25, 62b
Museum of Modern Art, New York, gift of R.E.M. 74
National Film Archive, (c) Paramount Pictures (UK) 104, (c) Turner Home Entertainment 107
Olivetti SpA. 91b, 133, 134ab
Popperfoto 89b, 98b, photo Westwood 9, photo Saidman 41
Archivio Rassegna 39, 76, 77, 78, 81, 82ab, 83
Courtesy of the Trustees of the Science Museum 14b, 16, 18ab, 19ab, 20, 21, 22, 23, 37ab, 42, 43, 64/5
Smart Design Inc. 150
Sony (UK) Ltd. 149a, 152
Sotheby's, Amsterdam 48
Sottsass Associati 103, 129
Team Concepts Electronics Ltd. 149b
Topham Picture Library 98a, 128, 131a
Vestry House Museum 30bc
Courtesy of the Trustees of the Victoria & Albert Museum 35
Zanotta SpA. 102

INDEX